E. Bretschneider

## On the Study and Value of Chinese Botanical Works

With notes on the history of plants and geographical botany from Chinese

sources

E. Bretschneider

**On the Study and Value of Chinese Botanical Works**
*With notes on the history of plants and geographical botany from Chinese sources*

ISBN/EAN: 9783337302825

Printed in Europe, USA, Canada, Australia, Japan

Cover: Foto ©berggeist007 / pixelio.de

More available books at **www.hansebooks.com**

ON THE STUDY AND VALUE OF CHINESE BOTANICAL
WORKS, WITH NOTES ON THE HISTORY OF
PLANTS AND GEOGRAPHICAL BOTANY
FROM CHINESE SOURCES.

BY

# E. BRETSCHNEIDER, M. D.

PHYSICIAN OF THE RUSSIAN

LEGATION AT PEKING.

ILLUSTRATED WITH 8 CHINESE WOOD CUTS.

Printed by Rozario, Marcal & Co., Foochow.

# PREFACE.

In presenting these pages to the learned world, I consider it my duty to confess, that I am neither a Sinologue nor Botanist, my knowledge in Chinese as well as in Botany being very limited. But living in the Chinese Metropolis five years, I was encouraged by the favourable conditions in which I found myself, to make some inquiries into Chinese plants and to venture on the publication of these notes on Chinese Botany. Every body will admit, I think, that some questions regarding Chinese plants can be more easily decided by men, living in China, by direct observation and information taken directly from the natives,—than in Europe by eminent savants, who have not been in China and must base their views, for the most part upon accounts given by travellers, which are not always exempt from errors, and upon translations from Chinese works, made by sinologues, who know little or nothing about Botany. I beg therefore to be excused if I have attempted sometimes to contradict some views of well known scholars. I implore indulgence for any errors which I may myself have committed. I have at least always endeavoured to adduce the sources whence I derived my information and prosecuted my enquiries in order to afford an opportunity for correcting or confirming my views. Although I had the advantage of having access to the splendid library of the Russian Ecclesiastical Mission at Peking, where are to be found not only all Chinese works of importance, but also most European books relating to China,—the reader will observe the want of some special works on Botany, indispensable in the treatment of botanical questions. But such works can only be met with in the great European libraries.—

As my notes have been written for Sinologues as well as for Botanists, I have endeavoured to be intelligible to both, and especially to the latter, by explanations of the Chinese characters, which occur therein. I would take advantage of this opportunity to observe, that Chinese names of plants should not be considered from the same point of view as names in other oriental languages, which can be transcribed easily and unmistakably by our letters. The Chinese language does not possess more than 400 words or monosyllabic sounds, distinguishable by an European ear. But as the Chinese characters (or hieroglyphs) are very numerous, each sound relates to a great number of characters of very different meaning. I will quote an example taken from the Chinese nomenclature of plants.

李 is a Plum,　　栗 a Chestnut,　　藜 a kind of Vegetable,
梨 a Pear,　　櫟 a kind of Oak,　　蘺 a kind of Garlic,

All these characters are pronounced by a sound, which must be rendered *Li* by European letters. In addition to this the Chinese characters, used by almost all peoples of Eastern Asia, are pronounced in a very different manner, not only by these different peoples, but even in different parts of China.\* Finally, European writers, ignorant of the Chinese language, frequently render Chinese names of plants very incorrectly and distortedly. This may suffice to prove, that it is completely useless and unintelligible to write the Chinese names of plants in European books, without the Chinese characters. The Chinese language is one suited more for the eye, than the ear. Therefore, in quoting Chinese names of plants, ambiguities can only be avoided by the addition of the Chinese characters.

---

\* 金橘 is the well-known little Kum-kwat orange (a variety of Citrus japonica.) The Chinese characters, meaning "Golden Orange" are pronounced *Kum-kwat* in the Southern dialect, but *Kin-kü* in the Mandarin dialect.

In transcribing the Chinese sounds by our letters, I have attempted to render them in the "Kuan-hua" or Mandarin dialect, the official language of the whole Empire, and which is at the same time the dialect of the Pekinese. With a few changes, I have adopted the mode of spelling in Mr. Wade's Peking Syllabary (but without tone marks). As is known, the Sinologues of each nation have a different system of transcribing the Chinese sounds, and each considers his mode as the best. But as it is impossible to render exactly Chinese sounds by any European letters, just as it is impossible for an European to pronounce Chinese sounds like a native, † this is a vain dispute. In my opinion the best mode of writing Chinese sounds is that, which requires the fewest letters. From this point of view I must declare the English language, so rich in useless letters, as not at all suitable. There are Chinese sounds, for the transcription of which Morrison (Dictionary) needs five letters, whilst by German or Russian spelling they can be rendered by two. For instance 七 *Tseih* (Morrison) can be written in German as well as in Russian, by two letters.—The English *ch, sh, yew, ye* can be rendered in Russian each by one letter. In addition to this the Chinese have sounds, which can only be represented exactly by Russian spelling. The other European languages, for instance, do not possess letters, like the Russian, for transcribing such characters as 子 and 四 *(tsze* and *sze* of the English Sinologues.) I will not however maintain, that the Russian language is the best for spelling Chinese sounds, for it cannot transcribe all Chinese sounds. It is for instance impossible to write with Russian letters such sounds as *shang, tung, fang, ting* &c., for the Russian language does not possess the nasel *ng*.

E. B.

Peking, December 17th, 1870.

† I must however except the Europeans born in China and who have spoken Chinese from their youth These acquire perfectly the Chinese pronounciation.

The object of the following pages is to show in what manner the Chinese treat natural science and specially botany, and what advantage can be drawn by European savants from the study of Chinese botanical works. As the principal works of the Chinese on Natural History have properly a medical bearing, I shall in criticising those works, occasionally make a few remarks also on Chinese therapeutics. Finally, I intend to give some characteristic specimens of Chinese descriptions of plants and add also a few Chinese woodcuts.

The Chinese knowledge of plants is as old as their medicine and agriculture and dated from remote antiquity. In ancient Greece the first botanists were the gatherers of medical plants. In the same manner the ancient Chinese got acquainted with plants for the most part in their application to medical purposes. There is a tradition among the Chinese, that the Emperor Shên-nung, who reigned about 2700 B. C., is the Father of Agriculture and Medicine. He sowed first the five kinds of corns (v. i.) and put together the first treatise on medical plants in a work known as 神農本草經 *Shên-nung-pên-ts'ao-king*, Classical herbal of Shên-nung (generally quoted by Chinese authors under the name *pên-king*), which became the foundation of all the later works on the same subject. This is a small work of 3 chapters, and enumerates according to the Pên-ts'ao in all, 347 medicines. 239 of them are plants, for the most part wild growing plants, but only very few cultivated ones. It follows from the accounts given by Li-shi-chên of this work (Preface of the Pên-ts'ao-kang-mu), that at first it existed only in verbal tradition. It is not known at what time the Shên-nung-pên-ts'ao was first written down, but there can be no doubt that it is one of the most ancient documents of Chinese materia medica.

Another very ancient work, which gives accounts of plants, known by the Chinese in ancient times, is the 爾雅 *Rh-ya*, a dictionary of terms used in Chinese ancient writings, which according to tradition has been handed down in part from the 12th century B. C. The greater part however is attributed to *Tsŭ-sia*, a disciple of Confucius. It is divided into 19 sections. The greater half of the work treats of natural objects. There is an enumeration of nearly 300 plants and as many animals of which also drawings are given. The Rh-ya is commented by 郭璞 *Ko-po* in the 4th century.

The first purely botanical work which appeared in China seems to be the 南方草木狀 *Nan-fang-ts'ao-mu-ch'uang* by 稽含 *Ki-han*, an author of the Tsin dynasty (265-419). It is divided into 4 divisions, herbs, trees, fruits and bamboos, and contains in toto the description of 79 plants of Southern China.

The Chinese works on materia medica and plants from the 6th to the 16th century are very numerous. The epoch of the T'ang (618-907) and the Sung (960-1280) especially was very productive in writers in this department. I cannot here enter into an enumeration of all their works. It would be useless, moreover, as I intend to speak of the well-known treatise on Chinese materia medica 本草綱目 *Pên-ts'ao-kang-mu*, for it is the type of all the Chinese productions of this class. 李時珍 *Li-shi-chên*, the celebrated author of the Pên.-ts'ao-kang-mu, a native of 蘄 州 *Ki-chou* in Hupeh has made extracts from upwards of 800 preceding authors. After having spent 30 years on the work, Li-shi-chên published it at the close of the 16th century. It can be said, that Li is the first and last critical writer on Chinese natural science and that he has never been rivaled by other authors. As has been already stated above, the greatest part of the work is purely medical, a specification of numerous prescriptions, of the pharmacological effect of the medicines and the complaints for which they are used. This part of the work is, I believe. without interest, not only for our naturalists, but also for medical students. The whole of the Chinese medical science is nonsense; their practice is for the most part not the result of experience. The Chinese have neither studied anatomy and the physiological functions of the human body, nor have they investigated, free from prejudices and superstition, the effect of their medicines. The art of healing in China is nearly in the same state now, as it was 46 centuries ago. The terms used in Chinese medicine to designate the action of medicines are quite as intelligible to the Europeans as to the Chinese physician. I need only cite some phrases, which occur in every Chinese book on medicine:

" All medicines, that are sweet belong to
" the element earth and effect the stomach ;
" all medicines, that are bitter belong to the
" element fire, which enters the heart." etc.

" All medicines, on account of their prop-
" erties, that are cold, hot, warm, and cool,
" belong to the *yang*, or male energy in na-
" ture, while their tastes, as sour, bitter,
" sweet, acid, and salt, belong to the *yin*, or
" female energy."

" The upper and lower, the internal and
" the external parts of medicinal plants have
" each their correspondent effects on the
" human system. The peel or bark has in-
" fluence over the flesh and skin ; the heart
" (pith) operates on the viscera etc. The
" upper half of the roots of medicinal plants
" has the properties of ascending the system,
" while the lower half has that of descend-
" ing."

It may be said, that there is in China no substance, not absolutely poisonous, no matter of what origin, which is not used by the Chinese as medicine. Often the most disgusting things are prescribed by Chinese physicians. A famous medicine is, for instance, the 人 中 黃 *Jen-chung-huang* (man's middle yellow) prepared from Liquorice, which has been placed for some weeks in human excrement. I once saw this abominable medicine prescribed in typhus fever, together with fifteen other drugs.

Luckily for the Chinese people, the native physicians do not like to prescribe efficacious medicaments ; their medicines are for the most part indifferent, and the method of preservation in the shops is such an unsuitable one, that many drugs lose their efficacy. In the neighbourhood of Peking, there is to be found an abundance of excellent Peppermint, 薄 荷 *Po-hò,* containing much more volatile oil, than our European plants. But the exsiccated plant, obtained from the Chinese druggists differs scarcely from hay. It is likewise difficult to find in the Chinese apothecary-shops Rhubarb of good quality. Although the best Rhubarb in European commerce is that brought from China*, that

* As is known, *the best Rhubarb* is that called *Moscovite* Rhubarb. In reality it came from China through Russia by way of Kiakhta, since the year 1767. Formerly the Russian government established a commission of experts in Kiakhta in order to examine carefully the drug carried by Chinese merchants. The completely faultless roots only were selected, whilst the inferior pieces were burned. The import of other Rhubarb was prohibited and only the crown Rhubarb was admitted for use in the Russian apothecary-shops. But some years ago the Russian government abolished this commission, and the apothecaries themselves now must look after their supply of Rhubarb. A great part of the Rhubarb used in Europe comes from the Chinese province Ssü-chuan or from the Himalaya. These are inferior sorts. The plants which furnish the Indian or Himalayan Rhubarb are described by our botanists as Rheum Emodi Rh. Webbianum etc. But regarding the Chinese Rhubarb and especially the Rhubarb, which is brought to Kiakhta, up to the present time neither the plants, which yield these drugs, have been seen by Europeans, or are their native countries known with certainty. The Kiakhta Rhubarb differs from other sorts in the drug having the form of a horse's hoof. The Chinese merchants, who bring the Rhubarb to Kiakhta, know nothing about the plant; they are acquainted with the roots only. I was informed by a Chinese Mandarin from Kan-su, that this Rhubarb thrives only on certain mountains in Kookonor and Kan-su and that this region is inhabited by wild tribes, completely independent of the Chinese government. They collect and prepare the Rhubarb roots and sell them to the Chinese at a fixed neutral place, whither purchasers and sellers repair armed. At first Rhubarb was brought directly from Kan-su to Kiakhta by Turkistan merchants, who in European writings are erroneously called Bukhars. But in later times, the Chinese of Shan-si, who up to the year 1861 managed also the tea commerce seized this commerce. The Russian name for Rhubarb is *rewen;* the Persians, who obtain their Rhubarb through Bukhara call it in a similar manner, *ríwend.* The Chinese name of

Rhubarb is 大 黃 *Ta-huang* (great yellow). It was known by the Chinese from remote times and is treated of in the herbal of the Emperor Shên-nung under the names *ta-huang* and

黃 艮 *Huang-liang* (yellow excellent). The Chinese distinguish a great number of kinds of the drug. Lately a collection of the best kinds, according to the Chinese, was made in Peking and sent for examination to St. Petersburg. The result was, that none of these specimens could rival the selected Kiakhta Rhubarb.

used commonly in China is worm eaten and of little value.

The pharmaceutical part of the Pên-ts'ao and the therapeutics of the Chinese can only interest us as a curiosity, as far at least as their medical views permit us to judge of the state of their culture. Our materia medica can learn nothing more from the Pên-ts'ao. It is undeniable, that the Chinese possess several very good medicaments, especially stomachics, amara &c., but we possess either the same plants, or others of a similar action. What is profitable among the Chinese medicaments, such as Rhubarb, Camphor, Star Anise, and I may also mention the Tea, we have incorporated many years ago into our pharmacopœas. The celebrated *Ginseng*, Panax Ginseng, 人參 *Jen-shên*, of the Chinese, enjoyed in Europe also a great reputation for some time, but it has been long ago rejected as an expensive and needless medicine.

The whole work of Li-shi-chên embraces 52 Chapters, and is divided into several sections. In this work, inorganic substances are arranged under the heads water, fire (Chapter 5-6), earth, metals, gems and stone (Chapter 7-11). Plants are comprised in 26 chapters (12-37); Zoology in 14 chapters (39-52).

According to the natural system of Li-shi-chên the plants are arranged under five divisions or 部 *pu*. These are still further divided into families or 類 *lei* which comprise the species or 種 *chung*.

## I. 草部 *Ts'ao-pu*, HERBS.

1. 山草 *Shan-ts'ao*, hill plants, such as grow wild. Ginseng, Liquorice, Polygala, Orobanche, Salvia, Scutellaria, Turnefortia Arguzina, Platycodon, Gentiana, Convallaria, Uvularia, Narcissus &c.,--78 species.

2. 芳草 *Fang-ts'ao*, fragrant plants. Levisticum, Paeonia Mutan, P. albiflora. Chavica Betel, Nutmeg, Turmeric, Amomum, Galanga, Nardostachys, Putchuk (?), Jasminum Sambac, J. officinale, Lophanthus, Mentha piperita, &c.,—60 species.

3. 隰草 *Shi-ts'ao*, marshy plants.— Chrysanthemum, Aster, different species of Artemisia, Carthamus tinctorius, Saffron, Boehmeria nivea, Xanthium strumarium, Arundo phragmites, Plantain, Ephedra, Juncus, Althaea, Hibiscus Abelmoschus, Kochia scoparia, Dianthus, Plantago, Silene, Polygonum, different Indigo-plants, Carduus, Sedum, Siegesbeckia, Tribulus terrestris,

Rehmannia glutinosa, Ophiopogon, Physalis Alkekengi, Inula, Iris, Arctium Lappa, Picris, Verbena, Sida tiliaefolia, Gnaphalium, Bidens, Cock's comb, Equisetum, Jasminum nudiflorum.—137 species.

4. 毒草 *Tu-ts'ao*, poisonous plants. Rhubarb, Phytolacca, Pardanthus, Ranunculus, Arum macrourum, Aconitum, Euphorbia, Ricinus, Veratrum, Datura, Balsamine.—54 species.

5. 蔓草 *Man-ts'ao*, scandent plants. Cuscuta, Convolvulus, monthly Rose, Pachyrhizus, Smilax sina, Rubia, Akebia quinata, Thladiantha dubia, Bignonia, Ficus stipulata, Hedera, Quisqualis, Muretia Cochinchin, Aristolochia, Kadsura, Melanthium, Roxburgia, Pharbitis Nil, Lonicera sinensis, Humulus.—113 species.

6. 水草 *Shui-ts'ao*, waterplants. Alisma, Acorus, Typha, Lemna, Marsilea Limnanthemum, Laminaria Saccharina, Myriophyllum.—29 species.

7. 石草 *Shi-ts'ao*, plants growing on rocks or in stony places. Dendrobium, Oxalis, Saxifraga, Fern, Sempervivum, Sedum.—27 species.

8. 苔 *Tai* family of mosses. Lichen, Lycoperdon, Lycopodium.—18 species.

9. 雜草 Miscellaneous plants not used in medicine.—162 species.

## II. 穀部 *Ku-pu*. GRAINS.

1. 麻麥稻類 *ma-mai-tuo-lei*, Hemp, Barley, Wheat, Buckwheat, Sesam, Rice.—9 species.

2. 稷粟類 *tsi-su-lei*. Millet, Sorgho Maize, Opium, Poppy, Coix lacryma.—17 species.

3. 菽豆 *Shu-tou*, leguminous plants. Sojabean, Dolichos, Phaseolus, Vicia Faba, Pisum sativum, Lablab.—13 species.

## III. 菜部 *Ts'ai-pu*. KITCHEN HERBS.

1. 葷辛類 *Sün-sin-lei*, pungent plants. Leeks, Garlic, Onion, Mustard, Cabbage, Ginger, Anthemis tintoria, Carrot, Radish, Persil, Star-Anise, Fennel.—38 species.

2. 柔滑類 *Jou-hua-lei*, soft and mucilaginous plants. Spinage, Amaranthus Blitum, Medicago sativa, Purslane, Dandelion, Yamsroot, Sweet Potato, Taro, Lilium tigrinum, Bamboo sprouts, Basella rubra, Lactuca, Beet, Chenopodium.—46 species.

3. 蓏菜 *Lo-ts'ai*, vegetables producing fruits on the ground. Brinjal, Lagenaria, Benincasa cerifera, Trichosanthes anguinea, Momordica Charantia, Gourds.—12 species.

4. 水菜 *Shui-ts'ai*, aquatic vegetables. Fuci, Algae &c.—6 species.

5. 芝栭 *Chi-rh.* Mushrooms.—31 species.

## IV. 果 部 *Kwo-pu.* FRUITS.

1. 五果 *Wu-k:o*, the five fruits, cultivated or garden fruits. Different sort of Plums, Apricot, Peach, Chestnut, Jujube, Shorea robusta.—16 species.

2. 山果 *Shan-kuo*, wild or mountain fruits. Pear, Apple, Quince, Crataegus pinnatifida, Diospyrus Kaki, various kinds Oranges, Lemon, Pampelmoose, Medlar, Myrica sapida, Cherry, Salisburia adiantifolia, Hazelnut, Oaks, Pomegranate, Walnut.—36 species.

3. 夷果 *I-kuo*, foreign fruits.* Nephelium Litchi, N. Longan, Canarium album, C. pimela, Xanthoceras sorbifolia, Hovenia dulcis, Fig, Glyptostrobus heterophyllus, Torreya nucifera, Averrhoa Carambola, various Palms, Phoenix dactylifera, Areca Catchu, Cocoanut, Jackfruit.—40 species.

4. 味類 *Wei-lei*, aromatics. Various species of Xanthoxylon, Pepper, Cubebs, Rhus semialata, Thea Chinensis.—17 species.

5. 蓏類 *Lo-lei*, plants producing their fruit on the ground. Melons, Water Melon, Grapes, Sugar cane.—10 species.

6. 水果 *Shui-kuo*, aquatic fruits. Nelumbium speciosum, Euryale ferox, Sagittaria, Trapa bicornis, Scirpus tuberosus.—6 species.

7. Fruits not used in medicine.—Spondias, Cookia punctata.—22 species.

## V. 木 部 *Mu-pu.* TREES.

1. 香木 *Siang-mu*, odoriferous woods. Thuja, Pine, Cunninghamia, Cassia, Magnolia, Alöexylon, Cloves, Myrrh, Sandalwood, Camphor, Borneo Camphor, Liquidambar, Benjamin, Dragon's blood, Assafoetida, Olibanum, Sticklack.—41 species.

* It is singular, that Li-shi-chên entitles this chapter foreign fruits. Although there some foreign fruits are treated of, as the Date palm, the Jackfruit etc. most of the described fruits, however are, without any doubt, indigenous in China, grow exclusively in China and are not found elsewhere.

2. 喬木 *Kiao-mu*, tall stemmed trees. Varnish tree, Tallow tree, Croton, Elaeococca verrucosa, Sterculia, Ailanthus glandulosa, Cedrela sinensis, Sapindus, Pterocarpus flavus, Abrus precatorius, Melia azedarach, Sophora japonica, Gleditshia sinensis, Diospyrus ebenus, Rosewood, Acacia Julibrissin, Catalpa, Chamaerops Fortuni, Tamarix, Populus, Ulmus, Salix.—60 species.

3. 灌木 *Kuan-mu*, luxuriant growing trees. Mulberry, Broussonetia papyrifera, Gardenia florida, Ligustrum, Lycium Chinense, Chimonanthus fragrans, Hibiscus syriacus, H. Rosa sinensis, Cotton, Bombax, Cercis sinensis, Camellia.—53 species.

4. 寓木 *Yü-mu*, parasitic plants. Pachyma, Viscum.—13 species.

5. 包木 *Pao-mu*, flexible plants. Bamboo-species.—4 species.

6. Miscellaneous species.—27.

The Pên-ts'ao describes in all 1195 plants. Dr. Williams (Middle Kingdom) counts only 1094 plants. But he over-looked the numerous additions, which are not mentioned in the index.

The Pên-ts'ao-kang-mu, being originally a materia medica, the plants described are properly only medicinal plants. But as the Chinese use almost every plant known to to them, as medicine, the Pên-ts'ao gives a complete record of the botany of the Chinese.

It is well known by all who have read Chinese books, how indistinctly they are written for the most part, and how confusedly separate and single ideas are thrown together. The Chinese are in complete ignorance of our system of punctuation. Few breaks are to be met with indicating the beginning of a new subject. Very often in a whole chapter, treating of several different things, no break can be found. This does not trouble the Chinese, for they pretend only to understand the single sentences. They are neither struck by an illogical combination of the sentences in their writings nor by contradictions. This reproach, however, falls least heavily upon the Pên-ts'ao, which can be consulted more easily than the other Chinese botanical works.

The Pun-ts'ao in treating of the several kinds of plants, animals, stones &c., follows in every case the same system. All the names of the natural objects described are written in large characters, the names of the authors or books are for the most part in brackets. Each article is divided into paragraphs. The first contains the name and the synonyms of the plant; the second,

釋名, an explanation of the names. The third, 集解, gives the botanical description. These three alone can interest us, for the bulky remainder is consecrated to pharmacological and therapeutical notices.

The Chinese names of plants consist of one character, but very often they are found by 2 or 3 characters. Ten of the Chinese radicals denote plants, and their combinations with other characters form the greatest part of the names of plants, used in Chinese books. These characters are:

艸 ts‘ao or 艹 Herb. (140).—1423 combinations. The most of them denote names of plants. F. i. 艾 ai, Artemisia.—茗 ming, the book's name of Tea.—茜 tsien, Rubia.

木 mu, Wood (75).—1232 combinations. The names of most trees are to be found under this radical.—柞 tso, a kind of Oak. —榛 chên, Hazelnut.

The radical characters 禾 hô Paddy, Corn, (115), 米 mi, Rice (119), 麥 mai Wheat (199) and 黍 shu, Millet (202) and their combinations form the names of most kinds of corn. F. i. 秈 sien, a kind of Rice, 粟 su, a kind of Millet, 稃 mou, Barley.

The radical 瓜 kua (97) and its compositions relates almost exclusively to several kinds of Cucumbers, Melons, Gourds etc., whilst the radical 豆 tou (151) is conservated to the leguminous plants and pulse.

The radical 麻 ma (202) denotes hemp and other textile plants; the radical 竹 chu Bamboo.

After having enumerated the different names of the plant, according to different authors, the Pên-ts‘ao gives an etymological explanation of the names. For most part each plant is denoted by a peculiar character. For instance the character for Diospyrus Kaki is 柿 shi, for Euryale ferox 芡 kien. The common character for Tea is 茶 ch‘a. The Jujube (Zizyphus) is denoted by the character 棗 tsao, which is formed by two characters 朿, denoting thorn. It is, as explained in the Pên-ts‘ao, on account of the prickled appearance of the tree. The

characters 蘆 Lu and 葦 Wei denote Arundo phragmites. The plants, which enjoy, on account of their utility, a great renown, have even peculiar characters for all parts of the plant. According to the Rh-ya (v. s.) the root of the Nenuphar (Nelumbium Speciosum) is called 藕 Ou, the leaves and the stalk together 荷 Hô, (Pên-ts‘ao], the stalk 茄 Kia,* the lower part of the stalk, being in the mud 蔤 Mi, the leaf 葭 Sia, the bud of the flower 菡 Tan, the seed with the spungy testa 蓮 lien, the white seed without the testa 的 ti, the cotyledons with the plumule within the seed 薏 i. As is known the common name of the Nenuphar is 蓮花 lien-hua, and the torus is called 蓮蓬 lien-p‘éng. The male plant of the hemp Cannabis sativa, 枲 si is designated by the character 枲 si, whilst the female (seed bearing) plant is 苴 tsü.

The characters which express the name often relate to the appearance of the plant, their properties &c., F. i. Physalis Alkekengi, the Winter Cherry is 紅姑娘 Hung-ku-niang, red girl, on account of the red leafy bladder, which encloses the ripe fruit.—Celosia cristata, Cock's comb, bears the same name in Chinese 鷄冠 Ki-kuan. —Arachis hypogaea, Ground-nut is called 落花生 Lo-hua-shéng (the blossoms fall down and grow), as the Greek word hypogaea also denotes the fruits growing seemingly in the ground. After the fall of the flower the fruit curves downwards and the pod ripens in the soil.—The Chimonanthus fragrans is called 臘梅 La-mei for its blossoms appear in the 12th month (la), the Jasminum nudiflorum 迎春花 Ying-ch‘un-hua (flowers which go to meet the spring,) on account of the early appearance of its blossoms in spring.—Lilium tigrinum bears the Chinese name 百合 Po-hô (hundred together), owing to the numerous scales, which form the bulb. This bulb is largely used as food in China.

There are in China a great number of cultivated plants, which have been introduced from other countries, especially from In-

* The same character, Kie, denotes Solanum Melongena, Brinjal.

dia and Centralasia. Regarding these plants and other foreign plants, the Chinese have often tried to render the foreign name by Chinese sounds, especially the Sanscrit (梵 *tan*) name. F. i. the 娑羅 *So-lo* is the Shorea robusta, in Sanscrit *Sal* or *Saul*, a tree native of India. Buddha is said to be dead under a Sal tree. For this reason the tree is also called 天師栗 *T'ien-shi-li* (Chestnut of the heavenly praeceptor). Pên-ts'ao XXIX 30. As there are in Peking no Sal trees, the Buddhist priests in the temples adore under the name of So-lo-shu, a splendid Northern tree, the Aesculus Chinensis (A. turbinata) which thrives also in Japan.—The Sanscrit name of Sandalwood (Santalum album) *dshandana* is rendered in the Pên-ts'ao (XXXIV 35.) by the sounds 旃檀 *Chan-t'an*. The common Chinese name is 檀香 *T'an-siang.*—The Jackfruit, Artocarpus integrifol, in Sanscrit *paramita* is called 波羅密 *Po-lo-mi* in Chinese.—A Chinese name for Saffron *(Ziaferan* in Persian) is 撒法郎 *Sa-fa-lang* * (Pên-tsao XV 42).

On the whole it can be said of the Pên-ts'ao, that the descriptions of plants therein are very unsatisfactory. We find statements of the native country, of the form, the colour of the blossoms, the time of blooming &c. These accounts are insufficient, because the Chinese in describing the parts of plants, have not a botanical terminology, but the blossoms, leaves, fruits &c., are described, in comparing them with the blossoms, leaves and fruits of other plants, which are often unknown to the reader. Besides these mentioned, there are also statements given about the utility of the plants for economical and industrial purposes. The descriptions consist for the most part of successive quotations of authors, whereby the same statements are several times repeated. Finally Li-shi-chên gives also his own opinion and generally it is the most reasonable one of all. A great many are accompanied with woodcuts, but these are so rude, that very seldom can any conclusion be drawn from them.

About the close of the Ming appeared another botanical work 羣芳譜 *Kün-fang-pu,* a herbarium in 30 books, compiled by 王象晉 *Wang-siang-ts'in.* A con-

siderably enlarged edition was published in 1708 with the title 廣羣芳譜 *Kuang-kün-fang-pu* in 100 books. It seems to be copied for the most part from the Pên-ts'ao, but there is also much new information drawn from ancient and more recent authors. The work has no illustrations, but its great superiority lies in the splendid type. The Pên-ts'ao is often inconvenient for reference, the paper and the impression being bad and the misprints numerous.

A review of the cultivated plants is also to be found in the 授時通考 *shou-shi-t'ung-k'ao,* an excellent work on agriculture, horticulture and the various industrial sciences, issued by order of the Emperor in 1742, in 78 books. The drawings are tolerably good. Our Sinologues have often made translations from this work.

The last treatise on Chinese botany, of any note, issued in 1848 is the 植物名實圖考 *chi-wu-ming-shi-t'u-k'ao* by 吳其濬 *wu-ki-siin,* a native of Honan. The work was written in Tai-yüan-fu in Shan-si and revised by 陸應穀 *lu-ying-ku,* a native of Yünnan. It contains 60 chapters. The one half of the work consists of a description (for the most part very confused) of the plants now known to the Chinese. The printing is very distinct. The other half includes nearly 1800 carefully executed drawings. Although here also many mistakes occur, this work is incomparably the best pictorial work of the Chinese of this class. The price at Peking is about $14.

These are about the most remarkable Chinese botanical works, and which render unnecessary, reference to the numerous other works in this department.

I have announced at the outset of this article my intention to treat of the value of Chinese botanical works. Judging from the above remarks some may suppose, that I intend to deny all scientific value to their works. It is true, the Chinese possess very little talent for observation and zeal for truth, the principal conditions for the naturalist. The Chinese style is inaccurate and often ambigous. In addition to this the Chinese have an inclination to the marvelous and their opinions are often very puerile. None of the Chinese treatises can be compared with the admirable works of the ancient Romans and Greeks, *Plinius, Dioscorides* (both in the first century) &c. Nevertheless the Chinese works on natural science are very interesting, not only for sinologues, but also for our European natu-

---

* I must here correct my former statement (Notes and Queries IV p. 55.), that *yü kin siang* may be the Saffron.

ralists. One of the most interesting branches of botany, of more interest than systematic botany,* which usually consists only of dry monotonous description of plants, without any account of the relation of the plants to man—is geographical botany, and the history of the cultivated plants. The celebrated botanist *Mr. Alph. De Candolle* has already, in his remarkable work, Géographie Botanique, 1855, expressed his opinion, that the Chinese botanical works could throw light on some dubious questions in this department. He closes his work in the following terms: "L'ancienneté, en Chine et au Japon, de quelquesunes des races de plantes cultivées est curieuse, de même que la séparation du peuple Chinois d'avec les peuples de l'Inde, à une époque reculée, séparation qui se prouve par des cultures différentes et par des noms de plantes usuelles, absolument différents. J'ai senti à plusieurs reprises dans mes recherches combien l'étude des encyclopédies Chinoises et Japonaises pourrait rendre plus de services à l'histoire des espèces cultivées, laquelle à son tour est importante pour l'histoire des nations." Indeed, these works conceal accounts of interest; it is however very difficult "to fish out the pearls from the mud."† The pages of Notes and Queries have been much taken up with interesting discussions on this subject, especially on the introduction of certain cultivated plants into China. The Chinese authors agree in stating, that *Cotton* was introduced about the 9th or 10th century from Central Asia and Cochin China. In the same manner it can be proved from Chinese sources that *Maize* and *Tobacco* are not indigenous in China. Cf. Notes and Queries Vol. II No. 4, 5, Vol. I No. 6.

We can, I believe, assume with certainty, that all plants mentioned in the *Materia Medica* of the Emperor *Shên-nung*, in the Chinese classics (the *Shu-king*, the *Shi-king*, the *Chou-li*, the *Chun-tsiu* and other works of great antiquity*) and in the *Rh-ya* (v. s.) are indigenous in China and have not been introduced from other countries, for only about 120 B. C. the Chinese became acquainted with the distant countries of Asia, especially Western Asia. India, even then, they knew only by name. Before that time they had intercourse only with their nearest neighbours. It can also be said, that all plants designated in Chinese writings by *one* peculiar character, are indigenous.

I may be allowed to make here a few remarks on the products of the field and the garden in China and on the antiquity of their cultivation according to Chinese works. Although much has been written in Europe on Chinese agriculture, no details are to be found on the cereals cultivated by the Chinese. The following notes are for the most part taken from the Pên-ts'ao-kang-mu, † which quotes all the ancient works above mentioned.

Ssŭ-ma-ts'ien, the Herodotus of China, in his historical work 史 記 Shi-ki, written in the second century B. C., states that the Emperor Shen-nung 2700 B. C. sowed the five kinds of corn (藝 五 種) ‡ Cf.

* I do not wish, however, to be suspected of denying the great importance of systematic botany, the basis of all botanical science. The great confusion, however, which occurs in botanical nomenclature is to be deplored, for some botanists create unnecessarily new genera and species, which in reality do not exist. In this way the scientific synonyms of plants become very numerous and we are often embarrassed as to which name should be quoted. Sometimes it may be more intelligible to quote a popular indigenous name, than a scientific one. It would be very desirable, if the botanists of all nations would adopt the valuable work, just now published, of Bentham and Hooker, Genera plantarum, as a botanical code.

† It seems, that tho Chinese have a predilection for investigating the origin of natural objects. I need only cite the 格 致 鏡 原 in 100 books, published in 1735. In this work the origin and history of every subject is treated of in a long series of quotations from the native literature, ancient and modern; 16 books are dedicated to the investigation of the origin of the different plants, and represents therefore a kind of Chinese geographical botany. Another work in this department is the 毛 詩 名 物 圖 說. It contains an enumeration and description of all plants and animals mentioned in the Shi-king,

* The 書 經 Shu-king, "Book of History" compiled by Confucius (about 500 B. C.), the 詩 經 Shi-king, "Book of Odes," a collection of ballads used in ancient time, selected and arranged by Confucius.—The 春 秋 Ch'un-t'siu, Spring and Autumn Annals, also written by Confucius.—The 周 禮 Chou-li, "Ritual of the Chou dynasty," written about 1100 B. C. All these works have been translated into European languages. The 山 海 經 Shan-hai-king, "Hill and river classic" has nearly an equal antiquity.

† The abbreviation P in the following denotes the Pên-ts'ao kang-mu, the letters Ch. W. relate to the drawings in the Chi wu ming shi t'u k'ao.

‡ It is known, that at the vernal equinox the ceremony of ploughing the soil and sowing of the 5 kinds of corn are performed by the Emperor assisted by members of the boards. According to the 大 清 會 典 Ta-ts'ing-hui-tien, a description of the Chinese Government (Chap. 260 p. l.), where this ceremonial is described, the 5 corns sowed are 稻 Tao, (rice) 麥 Mai (wheat) 穀 Ku (Setaria italica) 黍 Shu (Panicum miliaceum) and 菽 Shu (Soja bean.) The Emperor sows the rice, the three princes and the members of the boards sow the remaining cereals. As I have been informed by the overseer of the Sien-nung-tan or temple of Agriculture in the Southern part of the Capital, where this ceremony is performed every year, the 5 cereals now used for this purpose are rice, wheat, Sorgho, Setaria italica, and the Soja bean.

Shi-ki Chap. 1. In later times the Chinese commentators agreed that here the following corns were meant:—

1 黍 *Shu*, 2 稷 *Tsi*, 3 菽 *Shu*, 4 麥 *Mai*, 5 稻 *Tao*. The Chou-li (Ritual of the Chou v. s.) states, Book V p. 5 (see also the French translation of Biot I p. 94), that vegetable and animal food must be combined ·n the following manner, enumerating 6 kinds of corn. The 稌 *Tu* (the same as 稻 *Tao*), rice suits with beef, the 黍 *Shu* with mutton, the 稷 *Tsi* with pork, the 粱 *Liang* with canine flesh, the 麥 *Mai* with the duck, the 苽 *Ku* with fish.

These cereals mentioned in the most ancient works, are up to this day cultivated in China. *v. p. 45*

黍 *Shu* (P. XXIII 3, Ch. W. I.) according to Dr. Williams, (Bridgman's Chrestomathy p. 449) this character denotes Sorgho. But at Peking *Panicum miliaceum* is called Shu and the description of this plant in the Pên-ts'ao suits more with Panicum. When hulled it is a roundish little corn of a pale yellow colour; when boiled it becomes very glutinous. The hulled corn is called 黃米 *Huang-mi*, (yellow corn) at Peking. From the Huang-mi the 黃酒 *Huang-tsiu*, yellow whisky is distilled.

稷 *Tsi* (P. XXIII I Ch. W. I.) The popular name in Peking is 藤子 *Mei-tsu*.

The Shu and the Mei-tsŭ are very similar in appearance, the plants as well as the corn. The difference consists in the Mei-tsŭ when boiled giving no gluten. This difference is also stated in the Pên-ts'ao. In addition to this the corn of the Mei-tsŭ is of a dark yellow colour. Prepared by boiling it is largely used as food (飯 *fan*) by the lower class. The Tsi or Mei-tsŭ is also a species of Panicum, allied to P. miliaceum. As I possess no specimens of our European P. miliaceum I am not sure whether the Shu or the Tsi agrees with the European plant. Bunge in his enumeration of Peking plants quotes the P. miliaceum.

粱 *Liang* (P. XXIII 7 Ch. W. I.) The popular name of the plant in Peking is 穀子 *Ku-tsŭ* the hulled corn is called 小米 *Siao-mi* (little corn.) It is of a yellow colour and much smaller than Shu-

tsŭ and Mei-tsŭ. This cereal is the *Setaria italica*. In Northern China, where the rice is dear it is largely cultivated and forms the principal food of the lower classes. The Pên-ts'ao explains, that this corn came first from 梁州 *Liang-chou* (an ancient country comprising a part of Shen-si and Ssŭ-chuan), hence the name. Other authors state, that the name is derived from the character 良 *Liang*, of like sound and meaning excellent. Therefore the Rh-ya writes 稂 *Liang*.

麥 *Mai*. Regarding the mai the Pên-ts'ao relates after the ancient dictionary 說文 *Shuo-wên* (published A. D. 100), that this corn is an excellent present, which came from heaven, therefore the character mai includes the character 來 *Lai*, ( to come.) The Shuo-wên states, that there are two kinds of mai, the 秾 *Lai* and the 麰 *Mou*, which characters often occur in the Chinese ancient books. The first denotes, as the Chinese authors explain, the 小麥 *Siao-mai*, or *Wheat* (P. XXII 17. Ch. W. I.), the second 大麥 *Ta-mai* or *Barley* (P. XXII 23. Ch. W. I.) Decandolle (l. e. p. 935) is therefore not right in assuming, that barley was not known by the ancient Chinese. The Pên-ts'ao states further, that the Sanserit name of wheat is 迦師錯 *Kia-shi-tsu*. Wheat and Barley are much cultivated in the neighbourhood of Peking. The common Chinese bread is made from wheaten meal, 白麪 Pai-mien.

稻 *Tao* is a general name for rice. The hulled corn is called 米 *Mi*, (P. XXII 29 Ch. W. I.) The Pên-ts'ao distinguishes the 糯 *No* or *glutinous rice*, which when boiled becomes glutinous and the 粳 *King*, (P. XXII 34) which yields no gluten, the 水稻 *Shui-tao* or *water rice*, and the 旱稻 *Han-tao* or *dry rice*, which does not require irrigation. In the neighbourhood of Peking, there is very little rice cultivated (on the banks of the river Hun); most of it comes from the southern provinces. The best rice in Peking is considered the 粳米 *King-mi*. It is very white.

The 白米 *Pai-mi* or common rice is an inferior sort. The glutinous rice* is called 江米 *Kiang-mi* in Peking, as it comes chiefly from Kiang-su.

What cultivated plant is meant by 菰 *Ku* I am not able to state. In the Pên-ts'ao it is called 菰米 *Ku-mi* (XXIII 15) and judging from the description therein it is a kind of corn cultivated in water. The Chi-wu-ming &c. takes no notice of this cereal.

菽 *Shu*. This name occurs in the Shi-king and in the Ch'un-ts'iu (v. s.) and was related in ancient times probably to the *Soja-bean†* (Soja (Glycine ) hispida. The Kuang-ya (4th century) says that the Shu and the 大豆 *Ta-tou* (great bean) are the same. The Pên-ts'ao (XXIV I and 8) states, that there are several kinds of Ta-tou, a black, a white and a yellow (so named after the colour of the seeds) and that from these beans 醬 *Tsiang* (Soja), 豆腐 *Tou-fu* (Bean-curd) and 豆油 *Tou-yu* (Bean-oil) are made. The drawing for *Ta-tou* in the Ch. W. I. represents the Soja hispida‡

All these plants, mentioned, are doubtless indigenous in China and cultivated there from remote times, according to the Chinese authors.

* As Mr. Billequin, an able Chemist in Peking, communicated kindly to me, the glutinous properties of this kind of rice are owing to the great quantity of Dextrine or Starch gummi contained in it. The common rice contains only 1 per cent Dextrine. (Cf. Payen, Substances alimentaires p. 265.)

† In the work of Loiseleur, "considération sur les céréales I p. 29, there is a translation from ancient Chinese works by M. Stan. Julien in which the character shu (one of the 5 cereals sowed by Emperor Shên-nung) is translated by "Fève" (Faba sativa, common Bean). With reason Decandolle, who refers to this translation (l. c. p. 956) is astonished that the common bean should be a native of China. As I will state below the common bean was introduced into China from Western Asia.

‡ At Peking two kinds of the Ta-tou are cultivated the 黃大豆 *Huang-ta-tou* (great yellow bean) and the 黑大豆 *Hei-ta-tou* (black great bean). The name great bean refers not to the seeds but to the whole plant, the Soja bean being an erect herb 3 to 4 feet high. The *Huang-ta-tou*, called also 毛豆 *Mao-tou* (hairy bean) is the true *Soja bean*, an erect hairy plant with trifoliate leaves, little axillare flowers, pendulous pods and white yellowish seeds of the size of a great pea, but a little oblong. This is the "Phaseolus japonicus erectus, siliquis Lupini, fructu Pisi majoris candido" described in Kaempfer Amoen. exot., the Dolichos Soja of Thunbeg.

The *Hei-ta-tou*, which resembles much the Soja bean, is also covered with red hairs the seeds are of the same size as the Huang-tou but black, I think it is a variety of the Soja bean. Both the yellow and the black bean are used for the same

The 薏苡 *Yi-yi*, (P. XXIII 17 Ch. W. I.) *Coix Lacryma*, Jobstears, is also a native of China for it is mentioned in the Shên-nung pen-ts'ao.

There is a plant called 稗 *Pai*, mentioned in Chinese books (P. XXIII 13 Ch. W. I.) and cultivated near Peking. It seems also long ago to have been cultivated, for the character pai occurs in the Shuo-wen (v. s.) This is the *Echinochloe Crus galli* of the botanists.

It cannot be decided from the Chinese authors, whether the Guinea corn *Sorghum vulgare*, now so extensively cultivated in Northern China as in Southern Europe, Africa, Western Asia and India, is indigenous to China. It is not mentioned in the Chinese classics.* The most ancient work, quoted by Li-shi-chên about the Sorgho is the 廣雅 *Kuang-ya*, written at the time of the Wei 386-558. The Chinese names for Sorgho are 蜀黍 *Shu-shu* (the first character denotes the province Ssŭ-ch'uan) 蘆粟 *Lu-su* (reed millet) 木稷 *Mu-tsi* (tree millet) (Kuang-ya), 高粱 *Kao-liang* (high millet.) The latter is the common name at Peking (P. XXIII 6. Ch. W. I.) In Peking where it grows plentifully it is employed chiefly for feeding horses and for distilling whisky, called 燒酒 *Shao-tsiu*.

Regarding the *Buckwheat* (Fagopyrum esculentum) 蕎麥 *Kiao-mai*, (P. XXII 26. Ch. W. I.) which is cultivated in Northern

purpose at Peking for making Soja, and Bean-curd. *Bean-curd* is one of the most important articles of food in China. It is prepared by macerating the above mentioned beans in water and milling them together with the water. The liquid pap is filtred. To this fluid is added gypsum in order to coagulate the Casein and also Chlormagnesium. The coagulated Casein or Bean-curd is a jelly-like appearance.

It is known, that Manchuria produces a large quantity of Beans (generally in the Reports on trade called Peas) from which by pressure *Bean-oil* or *Pea-oil* is made. Bean-oil is largely used in China for cooking and for lighting lamps. The *Bean-cakes* are exported to Swatow for purposes of manure in the Sugar plantations. New-chuang (in Manchuria) exports chiefly Bean-oil and Bean-cakes. I have not seen the Bean used in New-chuang for this purpose, but from the description of others it must be the Soja bean. Mr. Payen (l. c. 341) has examined leguminous fruits from China, which he calls *pois oleagineux* de la Chine and states, that they contain 13 per cent. oil, whilst our common leguminous seeds contain only 2 to 3 per cent. oil.

* Lacharme and Mohl in their translation of the Shi-king 1840 (the only one existing up to the present time) state that the *Kao-leang* or guinea corn is mentioned in the Shi-king (p. 51, 260, and 93). But in the Chinese text there is only the character *Liang* (v. s.) Setaria. The fancy of the translators has added the character Kao.

China, it is not certain, whether it is indigenous to China or introduced from Central Asia. The author, who first mentions buckwheat in China wrote during the Sung dynasty 960-1280. *

The character 麻 *Ma*, which now-a-days relates to all kinds of textile plants seems originally to have been used to designate the *common Hemp* (Cannabis sativa). As I have stated above, the Rh-ya notes a female ma, which furnishes only seeds and a male. This can only denote the Cannabis sativa, with the male and female flowers on distinct plants. The ma is mentioned in the Shu-king. The Pên-ts'ao calls it 大 麻 *Ta-ma* (great Hemp) P. XXII ii. Ch. W. I., and observes that the seeds of the ma are innoxious, whilst the leaves are poisonous. This agrees also with the Hemp.

Another textile plant mentioned in the Chou-li (Book XVI. translation of Biot I p. 379) and in the Shên-nung pen ts'ao, is the 葛 *Kŏ* (P. XVIII 42). It is according to the drawing in the Ch. W. XXII a twining Leguminosa; according to Hoffman and Schultes (Noms ind. d. plantesd. Japon et. d. l. Chine) *Pachyrrhizus Thunbergianus*.

In the same manner as the ancient Chinese enumerate 5 cardinal cereals, they distinguish also 5 garden fruits, 五 果 *Wu-kuo*. These fruits are, according to the Pên-ts'ao: 李 *li*, 杏 *sing*, 桃 *t'ao*, 栗 *li*, 棗 *tsao*, and as the Rh-ya, the Chou-li, Shi-king and other works of great antiquity mention them, there can no be doubt, that they are indigen-

ous. * The Sing, however, is not mentioned in the Chinese classics. *v. p. 45*

李 *Li* denotes plum. The Chinese have yet another term for plum. This is 梅 *Mei*, also an ancient name, which occurs often in the classics. This character comprises several kinds of edible plums and also very handsome ornamental flowers of the genus Prunus, with uneatable fruits. The 榆葉梅 *Yü-ye-mei* (plum with Elm leaves) is the *Prunus trichocarpa*. Its pink flowers appear early in February. Another beautiful ornamental shrub is the 紅 梅 *Hung-mei*, also a Prunus species with precocious flowers.—But the savoury fruit called 楊 梅 *Yang-mei* is furnished by *Myrica sapida*.

杏 *Sing*, as is well known, is the Apricot (Prunus armeniaca). This character can not be found as the name of a fruit either in the Shu-king or in the Shi-king, Chou-li &c. But the Shan-hai king states, that at the 霍 hills many Sing trees grow. In addition to this the name of the Apricot is represented by a peculiar character, which may prove, that it is indigenous in China. Our botanists assume, that the native country of the Apricot is the Caucasus and Western Asia.

桃 *T'ao* is the Peach, *Amygdalus persica*. Decandolle ( l. c. 889 ), believes that China is the native country of the Peach. He may be right.

---

* All the above mentioned cereals are cultivated in the plain of Peking. The Chinese records state, that at the time of the Yüan (Mongol) dynasty 1280-1368, the plain of Peking was hardly cultivated, it being used as pasture for Mongolian horses. Only since the Court of the Ming dynasty 1368-1644 which first resided at Nanking, was transferred to Peking (Emperor Yung lo 1403-1424) several Chinese cereals began to be sowed and at first only the Sorgho (Kao-liang) was cultivated.

It would, I think, not be without interest to give here a comparative list of the prices of the principal corns cultivated at Peking.

King-mi (best sort of rice)
| | | | |
|---|---|---|---|
| 1 catty (1⅓ ℔ English) .. | 560 | cash. |
| Kiang-mi (glutinous rice) 1 catty | 460 | ,, |
| Wheaten meal 1 catty .. .. | 360-420 | ,, |
| Pai-mi (common rice) 1 catty.. | 280 | ,, |
| Huang-mi (glutinous millet) 1 catty .. .. | 240 | ,, |
| Mei-tsŭ-mi (Panicum) 1 catty | 200 | ,, |
| Barley 1 catty .. .. | 195 | ,, |
| Siao-mi (Setaria ital) 1 catty | 160 | ,, |
| Kao-liang (Sorgho) 1 catty .. | 180 | ,, |
| Maize meal 1 catty .. .. | 140 | ,, |
| Buckwheat ,, .. .. | 130 | ,, |

1000 cash=7 pence=14½ cents.

* The Pên-t'sao mentions also a fruit 巴且杏 *Pa-tan-sing* (Pa-tan-Apricot) and gives the following description of it (P. XXIX 10.) This tree grows in the Western country of the Mohamedans. It resembles the Apricot but the leaves are smaller. The fruit has little flesh, the stone is like that of the plum, the husk is thin, the kernel is of a sweet taste like hazel-nuts. This description suits perfectly with the almond. As is known, the Almond tree grows everywhere in Western Asia. Its Persian name is *badam* and thus sounding nearly a) patan, Bunge states (Enum. plant, Chinae b reals that the Almond tree is cultivated in Peking. e can not confirm this statement. At least I have never seen Almond fruits in China. It is known, that the Almond tree (Amygdalus communis) as regards its flowers and leaves strongly resembles the Peach tree (Amygdalus persica), but the fruits are very different. As far as I know the Almond tree does not occur in China. What the Europeans call Almonds in China are the kernels of the Apricot 杏仁 *Sing-jen*. Therefore the Chinese compare the Almond tree with the Apricot. but not with the Peach. The Pên-ts'ao gives the name 忽鹿麻 *Hu-lu-ma* as a synonym of Pa-tan-sing, but at the end of the article *Wu-lou-tsu* (XXXI 21) it is stated, that Hu-lu-ma (the name for the dates) is not the same thing as Pa-tan-sing. I have adduced these statements, for Mr. Sampson has asserted (Notes and Queries III p. 150) that Pa-tan is the seaport *Pattan* in India, and that Pa-tan-sing is a Chin. synonym for the Date.

栗 *Li* is the chestnut, *Castanea vesca.*

棗 *Tsao,* the Jujube, *Zizyphus vulgaris* seems to be one of the most popular fruit trees of the Chinese. They enumerate a great number of varieties of the Jujube. The largest and best, known among the Europeans as Chinese Dates, come from Shan-tung. At Peking there are two varieties of Zizyphus vulgaris. *Z. vulginermis* 夏嫋棗 *Ka-ka-tsao* is a tree without prickles, and fruits as large as a plum. *Z. vulg. spinosus* is a small shrub, armed with numerous very sharp thorns. It grows everywhere. Bunge in his Enum. plant. Chinae bor. remarks rightly: "frequentissima et molestissima." The tops of the walls, which surround the Board of Punishments and other official buildings are covered with their dry branches. Lindley is wrong in stating (Treasury of Botany p. 220), that Caragana spinosa is used for this purpose. The fruit of this variety, known under the Chinese name 酸棗 *Suan-tsao* (sour Jujube) is of the size of a hazel-nut.

The 梨 *Li* or *Pear,* although indigenous and cultivated in China from remote times is not classed by the Chinese among the garden fruits, but is included in the *wild fruits.* Pears and Apples are generally insipid in China, but there is in Peking a small white Pear, 白梨 *Pai-li,* of excellent savour. It is also distinguished from other pears by its completely round apple-like shape. Large succulent pears come to Peking from Manchuria.

There are in Northern China several kinds of *Apples,* both wild and cultivated. The character 棠 *T'ang* relates generally to the *Crabapple* or sometimes to *Crataegus* and occurs in the Rh-ya. A very renowned kind of the T'ang is the 海棠 *Hai-t'ang.* *Pyrus baccifera* or a closely allied species, according to Hoffmann and Schultes, *P. spectabilis,* Ait. It is much cultivated as well on account of its beautiful blossoms as for the small fruits of the size of a hazelnut, which are made into sweet-meats. The Pên-ts'ao (XXX 5) explains the name Hait'ang (sea apple) by the fact, that this crabapple came first from 新羅 *Sin-lo,* an ancient country in Corea, beyond the gulf of Chili.*

---

* I must observe, however, that the name 秋海棠 *Ts'iu-hai-t'ang* (*Ts'iu*=autumn) is not used in China to designate a crabapple, but is applied to *Begonia discolor,* a much esteemed ornamental flower of Chinese gardens.

Our common *garden apple* is also cultivated in Northern China. There are several varieties, as 頻果 *Pin-kuo,* 沙果 *Sha-kuo.* Some kinds are of a large size, but their flavour is far inferior to apples in Europe.

Another fruit ranged by the Chinese among the wild fruits, and with more reason, than the cultivated pear and apple, is the 山樝 *Shan-cha* (P. XXX 12 Ch. W. XXXII). This is the *Crataegus pinnatifida,* Bge., growing abundantly in the hills to the West of Peking, where it attains a height of 20 to 30 feet. This shrub (or tree) is not cultivated, but the red fruit, much larger commonly than the fruit of Crataegus, and known by the common name 山裡紅 *Shan-li-hung* is collected at the hills. An excellent sweet meat 山樝糕 *Shan-cha-kao* is prepared from it. This fruit it mentioned in the Rh-ya.

The *Oranges,* of which there are a great variety in China, are also comprised by the Chinese authors among the wild fruits. There can be no doubt, that most of them are indigenous in China and cultivated from ancient times. This would be proved by each species or variety bearing not only a different name, but most of them being designated by peculiar characters and mentioned in the Shu-king, Rh-ya and other ancient works.

橘 *Kü,* is the most common name for oranges. This name occurs in the Shên-nung-pên-ts'ao and in the Shu-king P XXX. 25 Ch. W. XXXI.

金橘 *Kin-kü,* (gold orange), *Kum-kwat* Orange (Kum-kwat is the Southern pronunciation of Kin-kü) *Citrus Japonica.* The fruit is roundish and of the size of a small plum. Another variety with small oblong fruits, frequently cultivated at Peking, is called 金棗 *Kin-tsao* (golden Jujube), P. XXX 37 Ch. W. XXXI.

橙 *Ch'éng* (P. XXX 34 Ch. W. XXX). *Coolie Orange* (Bridgman's Chrest).

柑 *Kan* (P. XXX 32 Ch. W. XXXI.) *Coolie Mandarin Orange.* (Bridgman's Chrest).

櫾 *Yu* (P. XXX 35 Ch. W. XXXI), *Shaddock, Pumelo, Citrus decumana.* The best sorts of the Pumelo are brought to the Capital from Amoy. The Pumelo is mentioned in the Shu-king.

The common *Lemon tree* at Peking is frequently raised in a dwarf form in pots as an ornamental shrub and also on account of the lemons, which it produces and which do not differ from our European lemons. It is called 香桃 *Siang-t'ao* and may have been introduced. This name is not in Chinese books. The name 檸檬 *Ning-méng* given to the Lemon in Bridgman's Chrest. p. 443, is also not to be found in Chinese books. Perhaps by these sounds the Hindustan name of the Lemon, being *Nee-moo*, is rendered.

The 香櫞 *Siang-yüan* (P. XXX 36 Ch. W. XXXI.) is an acid ~~orange~~ *Citron* of great size cultivated at Peking. The peel is thick and very wrinkled. The Pên-tsao identifies the the Siang-yüan with the 佛手柑 *Fo-shou-kan* (Buddha's hand). P. XXX 36 Ch. W. XXXI. But these fruits are very different, as is stated also in the Kuang-kün-fang-pu LXV. p. 15 and 19. The Fo-shou-kan is the celebrated *Fingered Citron, Citrus sarcodactylus*, with its lobes separating into finger-like divisions. This division is not produced artificially. The Siang-yüan is first described in the Nan-fang-t'sao-mu-chuang (4th century), but the Fo-shou-kan is not there mentioned.

Among the trees, fruits and herbs, which are enumerated in the Rh-ya and the classics and which therefore must be indigenous in China, I would just mention the following: 槐 *Huai, Sophora japonica* (P. XXXVᵃ 31 Ch. W. XXXIII).—楝 *Lien*, Pride of India, *Melia Azedarach* (P. XXXVᵃ 28 Ch. W. XXXIII).—梧桐 *Wu-t'ung*, or 櫬 *Ch'ên, Sterculia platanifolia* (P. XXXVᵃ 25 Ch. W. XXXV).—桑 *Sang, Mulberry-tree* (P. XXXVI, Ch. W. XXXIII). The wild Mulberry-tree is called 檿 *Yen* in the Shu-king (Tribute of Yü).—攝攝 *Nie Nie*, or 楓 *Féng*, Liquidambar formosana (P. XXXIV. 43 Ch. W. XXXV).—漆 *Tsi*, the Varnish tree (P. XXXVᵃ 17 Ch. W. XXXIII) is mentioned in the materia medica of Emperor Shên-nung and in the Shu-king (Tribute of Yü). Dr. S. W. Williams states in his Chinese Commercial Guide, "The varnish used in making lackered ware is the resinous sap of one or more species of *Sumach* (Rhus or Vernix vernica) and the *Augia Sinensis* Lour., which grow best in Kiang-si, Che-kiang, Ssù-chuan. The natives however call only one sort *Tsi-shu*

or varnish tree." Lindley (Treasury of Botany p. 1210) states that *Calophyllum Augia* yields the Chinese Varnish. The representation of the Tsi-shu in the Ch. W. seems to relate to a Sumach.

The characters 樗 *Chu* and 栲 *Kao* (cf. Rh-ya and Shi-king) denotes the *Ailanthus glandulosa*, the Vernis du Japon of the French. The commentator of the Rh-ya ranges this tree among the varnish trees, as do the French. It grows very easily and rapidly and can be found everywhere in Peking; it thrives even between the bricks of the Peking walls.—A much celebrated tree of the Chinese is the 椿 *Ch'un, Cedrela sinensis*. The Pên-ts'ao states, that this is the same tree, mentioned in the Shu-king (Tribute of Yü) under the character 杶 *Ch'un* as being used for bows. The Cedrela sinensis grows also at Peking. The fragrant leaf-buds in spring are used by the Chinese for food. Now-a-days the Chinese apply the character Ch'un to both, the Ailanthus and the Cedrela, and distinguish the first as 臭椿 *Ch'ou-Ch'un* (stinking Ch'un), on account of the disagreeable odour of the flowers,—the Cedrela as 香椿 *Siang-ch'un*, (fragrant Ch'un). The large pinnate leaves of both trees are very like in appearance, but the botanist distinguishes them easily, by Ailanthus having two little glands near the basis of the leaflets. Good drawings of these trees can be found in the Ch. W. XXXV. See also P. XXXVᵃ 12.

I have already stated above, that the *Nenuphar* is mentioned in the Rh-ya. It is therefore indigenous in China as well as two other water-plants the *Trapa natans* and *Euryale ferox*. Trapa natans Caltrop bears the Chinese names 菱 *Ki* and 菱角 *Ling-küe*, (P. XXXIII 26 Ch. W. XXXII). Euryale ferox is called 芡 *Kien* or 鷄頭 *Ki-tou* (fowl's head) (P. XXXIII 27 Ch. W. XXXII). Mention is made of both in the Chou-li V. 35, Biot's translation I p. 108.

The character 芋 *Yü* denoting *Taro*, Arum esculentum (Colocasia antiquorum?) does not occur in the ancient classics, but the dictionary Shuo-wen (100 A. D.) describes this plant P. XXXII 31 Ch. W. IV.)

The Yams,Igname of the French, *Dioscoraea*, of which several species are cultivated in China (D. Batatas, D. alata, D. sativa,) is called 薯蕷 *Shu-yü* or 山藥 *Shan-yao* in Chinese books (P. XXVII 33 Ch. W. III). The latter name is in use at Peking. The Dioscorea is indigenous in China, for it is mentioned in the most ancient works, the materia medica of Emperor Shên-nung and the Shan-hai-king. Decandolle assumes (l. c. 819) that the Indian Archipelago is the native country of the cultivated species of Dioscoraca.

Decandolle conjectures also, (l. c. p. 821) that *Batatas edulis*, the Sweet Potato may be of American origin. But this plant was described in Chinese books a long time before the discovery of America in the Nan-fang-ts'ao-mu-ch'uang (3rd or 4th century). The Chinese authors state that the 甘藷 *Kan-chu* (the first character denotes sweet) is an important cultivated plant, the roots of which supply the place of corn in Southern China. The root is said to be of a reddish colour and as large as a goose egg. The flowers resemble the 旋花 *Süan-hua* (a species of Convolvulus according to the drawing in the Ch. W. XXII). This suits perfectly with the Sweet Potato as also with the fine drawing of the Sweet Potato in the Ch. W. VI. The Pen-ts'ao describes this plant XXVII 36. At Peking it is known as 白薯 *Pai-shu*, (white Potato). The charrter *Shu* seems to be applied to plants with tuberous edible roots.

*Phytolacca decandra*, the Virginian Poke, and *Phytolacca octandra* are assumed by the botanists as being of American origin (Decandolle l. c. 736). In Europe these plants appeared only 200 years ago. But Phytolacca is mentioned in the materia medica of Emperor Shên-nung under the name 商陸 *Shang-lu* and must therefore be indigenous in China. There can be no doubt, that Shang-lu is Phytolacca. See the good drawing in the Ch. W. XXIV. The description of Shang-lu in the P. XVII[a] 8 (poisonous plants) suits well with Phytolacca. I am not able to state, whether Phytolacca decandra or octandra be meant. Both are cultivated at Peking (Cf. Bunge, enumer. plant Chin. bor.) The Chinese use the thick fleshy root as medicine, as do also the aborigines in America.

The favoured garden flower 菊 *Kü*, *Chrysanthemum Chinense* was also known by the Chinese from remote times. See the Rh-ya and the materia medica of Shên-nung.

As regards the *Tea* (Thea sinensis, or Camellia Thea) the most renowned among Chinese cultivated plants and now well known by most peoples of the globe, there is no evidence to show, that the tea-shrub is other than indigenous to China. Lindley (Treasury of Botany) states however, that the only country, in which it has been found in a wild state, is Upper Assam, and adds, that a Japanese tradition, which ascribes its introduction into China to an Indian Buddhist priest, who visited that country in the 6th century, favours the supposition of its Indian origin. But this statement is not correct. It may be right as Dr. Williams states (Middle Kingdom II p. 127) that the general introduction of tea cultivation, does not date prior to the 8th or 9th century, but I must observe, that the Tea-shrub is mentioned in the ancient dictionary Rh-ya under the names 檟 *Kia* and 苦茶 *K'u-tu* (K'u= bitter) and a commentator of this work, who wrote in the 4th century A. D. explains, that this is a little tree, which resembles the 栀子 *Chi-tsü* (*Gardenia* species, the leaves of which resemble, indeed, the tea leaves). It grows in winter; (the leaves do not fall off). From the leaves can be made by boiling a hot beverage. Now (at the time of the commentator) the earliest gathering is called 茶 *Tu*, the latest 茗 *Ming*. Another name for the plant is 荈 *Chuan*. In the province of Ssû-chuan the people call the plant 苦茶 *K'u-tu*.—The Japanese tradition to which Mr. Lindley refers, can be found in Kaempfer's Japan. The Japanese legend says, that about A. D. 519, a Buddhist priest came to China, and in order to dedicate his soul entirely to God, he made a vow to pass the day and night in an uninterrupted and unbroken meditation. After many years of this continual watching he was at length so tired, that he fell asleep. On awaking the following morning he was so sorry, he had broken his vow, that he cut off both his eyelids and threw them on the ground. Returning to this place on the following day he observed, that each eyelid had become a shrub. This was the Tea-shrub, unknown until that time.—The Chinese seem not to know this legend. I am astonished, that the great botanist has based such a scientific view on this fable, and I would remark, that the Pên-ts'ao states expressly, that in China wild-growing tea can be found. The character 茶 *Ch'a*, now used to designate the tea-shrub, arose probably out of the ancient character 茶 *Tu*.

I would speak finally of a tree, the fruit of which for a long time has been known in Europe as *Chinese Star-anise*. The native country of the *Illicium anisatum*, which yields the Star-anise, has been the subject of many discussions by savants. Some tens of years ago Mr. de Vriese, a Dutch savant, asserted, that the native country of the Star-anise was not China, as usually supposed, but Japan.

(Tijdschrift voor Natuurlijke Geschiedenis en Physiologie 1834. Over de Ster-Anijs.) He was, however, refuted by M. Siebold, (Erwiederungen, über den Ursprung des Sternanises, 1837) who proved that the Japanese plant, Illicium religiosum does not yield the Star-anise of commerce, and that the latter, much used in Japanese medicine, was introduced into Japan from China or other countries. M. Hoffmann at last seeks to prove (Angaben aus Chines und Japan, Naturgesch von dem Illicium religiosum 1837) that the Star-anise is also not a native of China. He quotes the Pên-ts'ao and asserts, that there it is expressly stated, that the Star-anise is not indigenous to China, but is brought by foreign vessels. But the quotation of M. Hoffmann is wrong, for the Pên-ts'ao states on the contrary, that the Star-anise grows in the Southern provinces of China.

Under the name of 薥 香 Huai-siang or 茴 香 Hui-siang (siang=fragrant) the Pên-ts'ao describes at first (XXVI 62) a fragrant plant with leaves like hairs, little yellow flowers, which are arranged like an umbrella. The seeds resemble the barley. The best kind is said to come from Ning-sia (province of Kan-su.) This is without doubt the common Fennel (Foeniculum vulgare.) I have also examined the Hui-siang obtained from the Chinese Apothecary shops. After this description the Pên-ts'ao continues as follows:

There is yet another kind of Hui-siang which is brought by foreign vessels. The fruit is as large as the fruit of the 柏 Po (Thuja) and is divided into 8 corners, each of them containing a kernel like a bean, of a yellowish colour and a sweet taste like the common Hui-siang. This fruit is called Po-hui-siang (po=vessel) or 八 角 香 Pa-küĕ-siang (eight cornered Hui-siang.) This fruit grows in Kuang-tung and Kuang-si, namely in the departments situated near the foreign frontier (变 廣 諸 番 及 近 郡 皆 有) and that the best comes in foreign* vessels, wherefore it is called Vessel-star-anise. It can not therefore be called in question, that the Star-anise tree grows in China. Mr. Rondot (Commerce d'exportation de la Chine 1848 p. ii) states: "L'anis étoilé est porté à Canton par les jonques fokiénoises. Le plus renommé est celui de Tsiouen-tchou-fou. Il en vient également, mais en moindre quantité, du

* I think, the character 番 (foreign) here relates not to distant countries, but only to the Southern confines of China.

Kiang-si, du Yun-nan et même de quelques endroits du Koŭng-tong." Dr. Williams' (Commercial Guide) mentions Fokien, Japan and the Philippines as the native countries of the Star-anise. But Lindley (Treasury of Botany) says, that Star-anise (Illicium anisatum) is only found in China. I think Lindley is right. I do not know, whether our botanists possess in their herbariums a specimen of this plant. It seems not to occur in countries visited by foreigners.* The Star-anise is much used by the Chinese. It is therefore inconceivable how little information can be found in Chinese books about this tree. I looked over in the great Imperial Geography I-tung-chi, the enumeration of products of all departments in the provinces of Fukien, Kuangtung, Kuangsi, Kiangsi &c. Regarding the Star-anise there is only one statement, a quotation from the history of the Sung dynasty, that Star-anise is a tribute of the Southern part of 劍 州 Kien-chow (now Yen-ping-fu in Fukien.)† I have also searched for the same purpose in the special descriptions of those provinces (Kuang-tung Tung-chi, Kuang-si Tung-chi &c,) but without success.

In addition to the above statements the Pên-ts'ao describes the 小 茴 香 Siao-hui-siang, called also 蒔 蘿 Shi-lo, 慈 謀 勒 Tsŭ-mo-le (XXVI 65,) both foreign names according to Li-shi-chên. This is also a fragrant umbelliferous plant, the black seeds of which are used as medicine. The native country is said to be Po-ssŭ (Persia). I am not able to state from this description, whether this is the Anise (Pimpinella Anisum) as M. Hoffmann asserts. The Persian name of Anise is Anisun i rumi (rumi=Roman), the name of Fennel is badian or rasianeh. The drawing of the Shi-lo in the Ch. W. IV resembles the Fennel more than the Anise.

Having in the foregoing remarks examined the most important of the indigenous cultivated plants in China, I would now refer shortly to the plants introduced from other countries into China.

* I would be greatly obliged if any of the readers of the Recorder, residing in Southern China, and especially in Fukien, could give information about the districts, where the Star-anise grows.

† This may be an example of the manner, in which the I-tung-chi and other Chinese geographical works, issued by Imperial command in the last century, are got up. We err in supposing, that all the accounts of the several provinces and districts etc. are collected directly from the Chinese authorities of the respective countries. These works were compiled in Peking from the most ancient Chinese books. For instance the products in the Kuang-tung Tung-chi and Kuang-si Tung-chi etc. are enumerated and described for the most part, according to the Nan fang ts'ao mu chuang (y. s.) a book, which appeared 1500 years ago.

During the reign of the Emperor Wu-ti 140-86 B. C. the *Si-yü* (the countries of Central Asia were opened up by the Chinese armies, and China then first became acquainted with the far West of Asia. The celebrated General 張騫 *Chang-kien*, the conqueror of the Si-yü, advanced to 大宛 *Ta-wan* (Kokand) and still further to 大夏 *Ta-sia* (Bactria). After having been absent for 10 years, he returned to China and brought along with him many useful plants from Western Asia, which soon spread over the whole of China and are cultivated here up to the present time. The Pên-ts'ao mentions the following plants as being introduced from Western Asia by Chang-kien, but some of them were probably earlier known by the Chinese, and Chang-kien only introduced better varieties. 蠶豆 *Ts'an-tau* (ts'an denotes silkworm. The pods are said to resemble the silkworm) or 胡豆 *Hu-tau*.* This is the Faba sativa, common Bean, a native of Europe and Western Asia. (Cf. Decandolle l. c. 956) P. XXIV 20 Ch. W. I (a fine drawing). The Kidney bean is still much cultivated at Peking under the name of Ts'an-tau.

Chang-kien further brought from the West the 胡瓜 *Hu-kua* or 黃瓜 *Huang-kua*, the *Cucumber*, (P. XXVIII 14 Ch. W. IV), the 胡荽 *Hu-sui* or *Parsley* (Petroselinum sativum) P. XXVI 55 Ch. W. IV., the 苜蓿 *Mu-su, Lucerne* or Medicago sativa P. XXVII 8 Ch. W. III Cf. Notice sur la plante Mou-sou p. M. Skatschkoff and M. Pauthier, 1864. Decandolle (l. c. 838) says about the Lucerne: "Les Grecs et les Romains l'appelaient Médika, herba medica, parcequ'ils la regardaient comme apportée de Médie (Plin. XVIII C. 16).

The Pên-ts'ao states also, that the 紅籃花 *Hung-lan-hua* or 紅花 *Hung-hua* (red flower) was brought to China by Chang-kien. This is the *Safflower, Bastard Saffron* or Carthamus tinctorius, used in China as well as in Western Asia and Europe for dyeing red. P XV 40 Ch. W. XIV.

At the same time the Chinese were acquainted also with the *Saffron*,* according to the Pên-ts'ao. The Saffron, Crocus sativus, is therein described (XV 42) under the name 番紅花 *Fan-hung-hua* (foreign Safflower). As synonyms are given 洎夫藍 *Ki-fu-lan*† and 撒法郎 *Sa-fa-lang*. Without doubt by these sounds is rendered the Arabian or Persian name Ziaferan. The Pên-ts'ao states, that this plant grows in Thibet (Sifan), in the countries of the Mohametans (Hui-hui-ti) and in Arabia (T'ien-fang). At the time of the Yüan dynasty (1280-1368) they mixed the Sa-fa-lang with their food. (This custom is up to the present time, found in Persia, where the rice is mixed with Saffron). At Peking the Saffron is known by the name 西藏紅花 *Si-tsang-hung-hua* (Red flower from Thibet), but it is not cultivated here. It is, however known, that the Saffron now is extensively cultivated in other parts of China. The Saffron (Crocus sativus) and the Safflower (Carthamus tinctorius) belonging to two different families and classes of the natural system [Iridaceae (Monocotyledons) and Compositae (Dicotyledons)] have not the slightest resemblance. It is therefore strange, that almost all nations, like the Chinese, confound these plants. Decandolle (l. c. 858) says: "Je remarque une certaine confusion chez les Arabes entre le Safran et le Carthame, dont les fleurs donnent aussi une teinture jaune et qui est cultivé en Egypte, où le Safran ne l'est pas. Le nom du Carthame en Arabe est *quotom*, celui de la fleur cette plante *ósfour*. Le premier rappelle le nom hébreu et persan du Crocus, le second vient de sa couleur et de l'analogie avec le Safran. Le Carthame a reçu dans le commerce le nom de faux Safran ou Safranon. On voit dans les anciens auteurs et déja dans Pline, que des emplois analogues ont fait de tout temps rapprocher et désigner semblablement ces deux plantes."

The Chinese distinguish two kinds of *Garlic*, the 葫 *Hu* or 大蒜 *Ta-suan* (great Garlic) and the 蒜 *Suan* or 小蒜 *Siao-suan* (small Garlic). The first is said (P.

---

* If the character 胡 occurs in the name of a plant, it can be assumed, that the plant is of foreign origin and especially from Western Asia, for by 胡人 *Hu-jen* the ancient Chinese denoted the peoples of Western Asia. They explain, that the writing of the Hu-jen is not arranged in vertical columns as the Chinese, but runs from right to left.

* I would here mention an error I committed in my article on Chinese ancient geographical names in stating, that 鬱金香 *Yü-kin-siang* might be the Saffron. By this name probably the Sumbul, Sumbulus moshatus, is meant.

† The character Ki is probably a misprint and must be written 咱 *Tsa*.

XXVI 21) to have been introduced from Western Asia, whilst the smaller sort seems to be indigenous. The character, Suan occurs in the Rh-ya. It can therefore be assumed, that the Chinese from remote times stunk of Garlic as now a days. In Western Asia also, the Garlic is one of the indispensable vegetables among all classes of the people.

The Pên-ts'ao states also (XXII 1) that the Sesamum orientale 胡麻 Hu-ma* was brought by Chang-kien from Ta-wan (Kokand). But there is here a contradiction, for ' Li-shi-chên believes, that the 巨勝 Kü-shêng, mentioned in the materia medica of the Emperor, Shên-nung is the same plant as Hu-ma. Synonyms are 油麻 Yu-ma, (Yu=oil) on account of the oil obtained from the seeds and used for food, but the common name of Sesam in China is 芝麻 Chi-ma (the first character denotes properly a mushroom). A drawing of the Sesam is found in the Ch. W. I. p. i. The seeds and the oil of Sesam are as largely used for food in Western Asia as in China. The Persian name is kundshut.

The Chinese authors mention also some trees as being introduced into Chinese by Chang-kien.

The 胡桃 Hu-tao, or 核桃 Ho-tao (nut-peach (P. XXX 45.) Ch. W. XXXI.) was brought from 羌胡 Kiang-hu. Kiang was at the time of the Han dynasty the name for Thibet. Hu-tao is the Walnut-tree, Juglans regia. Li-shi-chên gives the Sanscrit name as 播羅師 Po-lo-shi.

The Pomegranate, Punica granatum, 安石榴 (P. XXX 22. Ch. W. XXXII.) was got from Western Asia. Li-shi-chên explains, that the name An-shi-liu is derived from the two countries An and Shi. Both were, at the time of Chang-kien, little realms dependent on 康 Kang (Samarcand). The character Liu is derived from 謹瘤 Chui-liu (Chui-liu denotes goitre, and the pomegranate resembles the goitre.) Hoffmann and Schultes (l. c.) state, that the pomegranate was brought to China from India.

It has been contested by Mr. Sampson (Notes and Queries III p. 50) that the Vine

葡萄 P'u-t'ao, was first introduced into China by Chang-kien from Western Asia, as the Chinese authors state (P. XXXIII 7. Ch. W. XXXII.) Mr. Sampson quotes from the Pên-ts'ao, which speaks of wild vine, growing in Shan-si. In fact Li-shi-chên describes such a plant under the name of 蘡薁 Ying-yü or 野葡萄 Ye-p'u-t'ao. But, I think we cannot, in every case, take à la lettre the character Ye, for the Chinese like much to set before the name of a cultivated plant the character Ye or 山 Shan (both denoting wild growing) in order to designate wild plants, which have some resemblance with the cultivated. In Peking a species of Ampelopsis is called Ye-p'u-t'ao. It is however very likely, that a wild growing vine exists in Northern China, but it cannot be proved, that the cultivated vine descends from it, and it is very dubious, whether it would be suitable for culture. We have therefore no ground to call in question the statements of the ancient Chinese, that the excellent vine, now growing plentifully in the whole of Northern China, was introduced from Western Asia, which is considered as the native country of our cultivated vine. Li-shi-chên, however, observes, that the vine was known by the Chinese before the time of Chang-kien, for it is mentioned in the materia medica of Emperor Shên-nung, and adds, that before the Han dynasty 隴西 Lung-si was known as a grape-growing country, but it was not introduced into China before 122 B. C. Before the time of the Han, Lung-si (in the province of Kan-su) did not belong to China.

Besides these cultivated plants introduced by Chang-kien, I will give a further list of plants brought from foreign countries to China, according to the Pên-ts'ao.

The common Pea (Pisum sativum,) 豌豆 Wan-tou (P. XXIV 18. a fine drawing in the Ch. W. II.) The Synonyms, as given in the Pên-ts'ao, 回回豆 Hui-hui-tou (Mohamedan pulse), 戎菽 Jung-shu (Western barbarian pulse) indicate a foreign origin. Li-shi-chên states, that the pea was introduced from 西胡 Si-hu (Western Asia.) In Bridgman's Chrestomathy p. 449 pea is called 荷蘭豆 Ho-lan-tou (Dutch pulse.) At Peking peas are not much cultivated.

The Spinage, Spinacia oleracea, 菠薐 Po-ling, 菠菜 Po-ts'ai (the common name

---

* In Northern China the name Hu-ma, however, is applied to the Lin, Linum usitatiss imum, which is cultivated in Shan-si and on the borders of Mongolia. Its introduction must be of more recent date, for the Pên-ts'ao does not speak of it. But in the Ch. W. II. p. 31 is a fine representation of the Lin, therein called Shan-si Hu-ma.

at Peking), 波斯草 *Po-ssŭ-ts'ao* (Persian herb) is said to come from Persia *(P. XXVII i. Ch. W. IV.)* The botanists consider Western Asia as the native country of the spinage and derive the names, Spinacia, Spinage, Spinat, épinards from the spinous seeds. But as the Persian name is *esfinadsh* our various names would seem more likely to be of Persian origin.

Decandolle says (l. e. 843) concerning *Lattuce*, Lactuca sativa: "rien ne prouve qu'elle fût connue en Chine de toute ancienneté, au contraire Loureiro dit, que les Européens l'avaient introduite à Macao." Decandolle believes, that it was introduced into China from Western Asia. He may be right. Although the Pên-ts'ao says nothing about the introduction, the 生菜 *Shéng-ts'ai* (the common name of Lattuce at Peking) or 白苣 *Pai-kü* seems not to be mentioned earlier than by the writers at the time of the T'ang (618-907.) *Cf. P. XXVII 17 Ch. W. IV.*

白芥 *Pai-kie, ( White Mustard,)* Sinapis alba was brought from Hu-jung (Western Asia.) *P. XXVI. 34.*

The *Watermelon,* 西瓜 *Si-kua* or 寒瓜 *Han-kua* (kua is a general term for cucurbitaceous plants, *Si,* denotes West, han, cold,) is, as the Chinese name denotes (Western melon) not indigenous. The Chinese authors state *(P. XXXII 6 Ch. W. XXXI)*, that the Chinese first got acquainted with this fruit at the time of the Wu-tai (the five small dynasties, which succeeded to the T'ang. 907-960.) It was brought from Central Asia. The Watermelon now thrives plentifully in Northern China, but the best come to the Capital from Hami.

The 絲瓜 *Ssŭ-kua.* Trichosanthes anguinea was introduced from Southern countries *(P. XXVIII 15 Ch. W. VI)* and for this reason it is also called 蠻瓜 *Mankua* (Cucumber of the Southern barbarians.) The character Ssŭ in the first name denotes silk thread. It is probably an allusion to the fringed blossoms. The Greek word Trichosanthes denoting "hairy flowers" is chosen for the same reason.

The *Carrot* (Daucus Carota) a favourite vegetable of the Chinese, was according to the Pên-ts'ao *(XXVI 57)* first brought from Western Asia to China at the time of the Yüan dynasty (1280-1368), hence the name 胡蘿蔔 *Hu-lo-po* (Western rape). A fine drawing of the Carrot is found in the *Ch. W. VI.*

Capsicum annum, *Cayenne pepper* is now a days much cultivated in China and was mentioned in the last century as a cultivated plant of Southern China by Loureiro. But it has not been noticed either in the Pên-ts'ao or in other Chinese books of more recent data. As the name denotes, the Cayenne pepper is a native of Southern America. Its Peking name is 辣椒 *La-tsiao* (pungent pepper), or 秦椒 *Ts'in-tsiao.* The drawing of the La-tsiao in the *Ch. W. VI. p. 20* does not agree with the Cayenne pepper, but seems to represent a native Capsicum with roundish fruits. Loureiro calls C. frutescens La-tsiao.

Some of our European writers have asserted, that the *Tobacco plant* is a native of China. Rondot (l. c.) mentions two indigenous Chinese species, *Nicotiana fruticosa* and *N. Chinensis.* But there is no proof in Chinese books, that Tobacco (as is known is a native of America) was known in China before the close of the 16th century. (Cf. Notes and Queries 1867 No. V.) Li-shichên, who wrote at that time, was not yet acquainted with the Tobacco. In the *Ch. W.* issued in the year 1848 a description and a drawing are given of the plant *(XXXIII)*, which is called 野煙 *Ye-yen* (wild smoke) or 菸 *Yen,* the latter, an ancient character, properly means stinking plant.

The *Potato* (Solanum tuberosum) likewise an American plant, the cultivation of which has spread over the greater part of

* The character 椒 *Tsiao* denotes properly the Chinese pepper, *Xanthoxylon.* The Pên-ts'ao notes several indigenous species of Tsiao (XXXII 1-9) namely 花椒 *Hua-tsiao,* 蜀椒 *Shu-tsiao,* 崖椒 *Ya-tsiao.* Judging from the drawing in the Ch. W. XXXIII most of them seem to be species of Xanthoxylon. The kind best known to Europeans is the Hua-tsiao (coloured pepper, on account of the red coloured fruits of an aromatic pungent taste.) But our botanists do not agree as regards the species to which this Xanthoxylon belongs.—Bunge (enum. plant Chin. bor.) describes the Hua-tsiao of Peking as Xanthoxylon nitidum. But Dr. Hance (Adversaria 1864) describes the same plant as a new species, Xanthoxylon Bungei. Hanbury (Chinese materia medica) asserts, that Hua-tsiao relates to Xanthoxylon alatum. The common *Black Pepper,* Piper nigrum bears the Chinese name 胡椒 *Hu-tsiao,* but does not grow in China. The Pên-ts'ao states that its Sanscrit name is 昧履支 *Mo-lü-chi.* According to Crawfurd (Dictionary of the Indian islands) the Sanscrit name of Pepper is *maricha.*

...he globe, has also found its way into China, but its cultivation here does not seem to be successful and supplies more the want of the European residents, than those of the aborigines, among whom it has not as yet found much favour. They prefer other indigenous tuberous plants, such as the Yam, the Sweet Potato, the Taro, Arrow-root &e. The Potato is cultivated in the neighbourhood of Peking principally in the sandy plain to the North of the Capital, but it does not grow plentifully. At Peking the potato is called 山藥豆 Shan-yao-tou, in Southern China, according to Bridgman's Chrestomathly 荷蘭薯 Ho-lan-shu, because the Dutch first brought it to China.

The *Ground nut* (Arachis hypogaea), *Lo-hua-shéng* (v. s.) is much cultivated throughout China as an article of food. The oil obtained from it is an important article of commerce. Crawfurd (l. e.) states that the Ground nut, extensively cultivated in the Archipelago was probably introduced from China or Japan. Brown (Bot. Congo p. 53) is of the same opinion. But I think, this plant has been introduced into China in the last century, for the Pên-ts'ao does not mention it. It is first described and represented in the *Ch. W. (XXXI)* under the names 落花生 Lo-hua-shéng and 番豆 Fan-tou (foreign bean.) In the descriptive part of the *Ch. W. Chap. XVI* it is stated, that the Lo-hua-shéng is not an indigenous plant, but came by way of sea from Southern countries. There it is said, that at the time of the Sung 960-1280 or the Yüan 1280-1368 棉花 Mien-hua, 番瓜 Fan-kua, 紅薯 Hung-shu and Lo-hua-shéng were first brought from the sea countries to Canton. *

I have already stated, that the *Maize*, a native of America has been introduced into China. Li-shi-chên was the first Chinese author,

* The author explains that Mien-hua (Cotton) at that time was called 吉貝 Ki-pei, the Hung-shu, 地瓜 Ti-kua (ground melon), the Lo-hua-shéng 地豆 Ti-tou (ground bean.)—Under the name of Fan-kua the Ch. W. describes and represents (XXXI) the *Carica papaya*. I am not able to state what plant by Hung-shu is meant. But, I think these statements are not very authentic. The author may be right that all the above mentioned plants were introduced into China, but he errs regarding the time of their introduction. The Carica papaya is a native of tropical America and could not be introduced into China before the discovery of America.

I would finally remark, that Decandolle (l. e. 968) is of opinion, that Arachis hypogaea is also of American origin.

who mentioned it at the close of the 16th century, under the name of 玉蜀黍 *Yü-shu-shu* (Jade Sorgho) P. XXIII 6. Ch. W. II. He states, that it was introduced from Central Asia. Now a days it is largely cultivated in China and bears in each province a different name (Cf. Notes and Queries 1867 No. 6). The Persian name of Maize is *ghendum i Mekkä* (wheat from Mecca.) That seems to prove, that the Maize, after having been brought to Europe spread over Asia from West to East. At Peking the Maize is called 玉米 *Yü-mi* (Jade corn.) Decandolle (l. e. p. 838) says: " M. Bunge, qui a traversé le nord de la Chine, jusqu' à Péking, m'a certifié n'avoir pas aperçu de Maïs." This statement is not correct. The Maize is abundantly cultivated in the neighbourhood of Peking and the bread baked from Maize forms one of the cheapest articles of food of the poor.)† I have asked about the Maize of several of the oldest men in Peking. They agree in stating, that as long as they can remember Maize was cultivated here. In addition to this a learned Chinese assured me, that in Chinese records it is said, that the cultivation of Maize near Peking dates from the end of the Ming dynasty 1380-1644.

Amongst our European cereals the *Oats* (Avena Sativa) is also to be found in the Chinese dominions, but it grows only in the mountainous countries of Shansi, in Southern Mongolia, and in Thibet. The Oats is mentioned in the History of the Tang dynasty 618-907 (Tang-shu Ch. 256 Article T'u-fan) under the name of 青稞 *t'sing-ko* as a product of Thibet. The Pên-ts'ao speaks of it briefly (Art. Ta-mai.) The Ch. W. I. p. 32 describes the Oats and gives an excellent drawing. Oats is known in Peking under the names 油麥 *Yu-mai* or 鈴鐺麥 *Ling-tang-mai* (ling-tang denotes little bells.) But it does not grow here. *p.44*

The *Rye* (Secale cereale) as far as I know, is nowhere cultivated in the Celestial Empire. M. Perny, however, in his Dictionnaire français-lat-chin, Art. production, mentions Rye (Seigle) as a product of China. I am very curious to know, where he found Rye.

I would finally mention, en passant, that in the gardens of the Emperor a splendid cereal plant is cultivated under the name of 御穀 *Yü-ku* (Imperial corn.) This is the *Penicillaria spicata*, with a *typha* like appearance. This plant is extensively cultivated in India under the name of *Bajri*. At Peking it is, as I have been informed, used for the Imperial table.

In the above mentioned botanical work, Nan-fang-ts'ao-mu-ch'uang (written in the 3rd

† The Maize is so cheap in Peking, that even the beggars enjoy from time to time the luxury of eating maizebread. As is known, the principal food of the beggars in China is the same as that, of which dogs are fed, and is often collected on the streets, where vegetable and animal remains of human repasts are thrown.

or 4th century) the renowned garden flower of the Chinese 末利 *Mo-li* is first spoken of. In the same work another garden flower 素馨 *Su-sing* or 即悉茗 *Ye-si-ming* is described (P. XIV*b* 66. Ch. W. XXX.) It is said that both were introduced from the countries of the *Hu-jen* (Western Asia) and from the Southern sea. These Chinese names refer the mo-li to *Jasminum Sambac* (a native of India and Western Asia), the *Ye-si-min* to *Jasminum officinale*. Its native country is said to be India; the Persian or Arabian name of the plant is Ya-semin. The Chinese name mo-li seems to be of Indian origin. In the ancient work of Büsching, Ostindien (II. p. 757) the Indian names of several kinds of Nyctanthes (Jasminum) are given and these names sound almost the same as mo-li. F. i. Nyctanthes auriculata Mullei.—N. Sambac Kudamalligei.—N. undulata, Malligei. *

These data which I have brought together from the Pên-ts'ao and other Chinese works, are intended only to show, that the study of Chinese botanical works is not without interest, as regards the decision of some botanical questions, especially of the native countries of cultivated plants. I have in the foregoing notices treated only of such plants, as are generally known and about which there can be no doubt as regards the identification of the Chinese names with the scientific ones. Now I will treat shortly of the difficulties, which the student of Chinese botanical works must overcome, in order to understand clearly the meaning of these writings.

If you take a Chinese botanical work in order to be informed about any plant, the first difficulty, that arises, is, to find out, where this plant is described. This is very often impossible, for the Chinese botanical works note from 5000 to 6000 names of plants, the synonyms of each plant being for the most part numerous. I have already stated, that the Chinese have nothing similar to the alphabetical index of our comprehensive works I have therefore been obliged in my studies to compose such an alphabetical index of all names of plants and synonyms, according to the sounds of the Chinese characters, not only of the Pên-ts'ao, but also of the drawings in the Chi-wu-ming &c. In this manner the description of the desired plant can be found in the shortest time.

It can not be said, that the style in the Pên-ts'ao presents difficulties. In describing the plants, the authors use for the most part always the same terms. The difficulties consist in the right interpretation of geographical names, which occur and in finding out at what

* The Mo-li-hua (Jasminum Sambac) is a favoured flower of the Chinese. In Peking there are special gardeners, who cultivate exclusively the Mo-li-hua. Every day in summer, the flower-buds are gathered before sun rise (without branches or leaves) and sold for the purpose of perfuming tea and snuff, and to adorn the head-dress of Chinese ladies.—The Ye-si-ming is not cultivated in Peking.

time the quoted authors wrote. It would be clear from the foregoing relations, that after having found the description of the plant in the Pên-ts'ao, the principal questions for solution are its native country and at what time it was first mentioned by the Chinese authors. The exact answer to these questions requires often the most extensive knowledge of the whole of Chinese science. Li-shi-chên has compiled the Pên-ts'ao from more than 800 ancient and more recent works, not only botanical, but also historical, geographical, philosophical, poetical &c. In quoting these works he never gives the whole title, but only one character of the author's name or one or more character of the name of the book. For instance, the character 頌 (properly denoting song,) which is met very often in consulting the Pên-ts'ao denotes the 圖經本草 written by 蘇頌 in the 11th century. It is almost in vain, that you ask your native teacher about such works. In the first chapter of the Pên-ts'ao, there is a list of most of the works quoted by Li-shi-chên, but only of 20 of them is the date of their issue given, with a short critique. The useful work of Mr. Wylie, Notes on Chinese Literature, 1867, although the best European work extant of Chinese Bibliography, is insufficient for our purposes. But few of the authors quoted in the Pên-ts'ao can there be found. The great catalogue of the Imperial library 四庫全書 總目 (1790) may contain information about all these works, but it is not easy to seek it in a Chinese work of 200 volumes. Therefore it is easily understood, that European savants, who translate articles from the Pên-ts'ao, as regards the quoted works, restrict themselves to the term: "a Chinese author says."

But, in addition it is necessary also to know at what time the quoted author wrote, for otherwise the native country of the plant can with difficulty be determined. At all times the Chinese endeavoured to complicate their science, so that they themselves do not find their way easily. They seem to place the value of their sciences in these complications. It is known, that from ancient times each of the Chinese Emperors bore, besides his dynastic name, a name for his reign, and this latter, was often changed. There are Emperors, who are registered in their Annals with from 10 to 15 names, each composed at least of two characters. The Chinese authors, in citing dates, refer only to these reign-names of the Emperors, which correspond to our ciphers to designate the date. In the same manner the Chinese liked at all times to change the names of their provinces, cities, &c. Almost every dynasty, after having succeeded to the throne, changed the names of the cities and also of the provinces of China. In this manner every city bore different names at different times. But as the number of the characters, used to designate geographical names is limited and as certain characters are particularly in favour for names of departments or districts, it happens very often that one geographical name relates to a great number of places. For instance 西平 *Si-p'ing* now-a-days the name of a district in the province Honan, was, at the time of the Post-Han, a country in Kan-su, at the time of the Wu a district in Kiangsi. During the T'ang dynasty Si-ping was in Yün-nan. The name of a province 江南 *Kiang-nan* (the meaning of the two characters is to-the-South-of-the-river) occurs often in the Pên-ts'ao. Here it does not mean the country to the South of the Yellow river so called by the present dynasty, (An-hui and Kiang-su,) but is to be understood as the Kiang-nan province of the T'ang dynasty to the South of the Yang-tse-kiang, comprising the greatest part of the modern province Fu-kien and Kiang-si. The name 南海 *Nan-hai* (South sea) refered in ancient times to Kuang-tung, but sometimes the Chinese also understand by this name the Indian Ocean and Archipelago. Cf. the historical maps in the Hai kuo-tu-chi, a work on historical geography, 1844. It is clear, that the greatest errors can be committed by the reader unacquainted with the time at which the respective Chinese authors wrote. In the year 1842 Biot published a useful work, Dictionaire des noms anciens et modernes des villes et arrondissements compris dans l'Empire Chinois. This work is translated from the 廣輿記 *Kuang-yü-ki*, a small geography of the Empire, and arranged in alphabetical order, but it proves to be insufficient to explain the geographical names, which occur in the Pên-ts'ao. The most complete work of Chinese geography, ancient and modern is, as is known the 大清一統志 *Ta-tsing-i-t'ung-chi*, or the *Geography of the Empire* of the present dynasty in 500 books. But it is impossible even for the Chinese to find out, without any data, a geographical name in this bulky work. The Chinese have no alphabetical index in their works, in order to facilitate reference to the book. There is however a Chinese geographical dictionary extant, which in some degree meets these wants, the 歷代地理志 *Li-tai-ti-li chi* in 20 books. This work is much more complete, than the Kuang-yü-ki and the geographical names, ancient and modern, are arranged according to a system under about 1600 characters. It is not quite easy to look for a name in this book, but it is at least not impossible to find it out. In disposing these 1600 characters after the radicals, this geographical dictionary can be made more practical for consultation.

In the Pên-ts'ao occur also very frequently names of ancient countries not included in China. These must be sought either in the histories of the various Dynasties, which always contain at the end notices of foreign countries,—or in the celebrated work of Ma-tuan-lin 文獻通考 *Wên-sien-t'ung-k'ao* (380 books), written in the 13th century. I need not observe, that you often seek in vain and that the demand for some explanation from the native scholars is equally fruitless.

Such are the difficulties to be overcome, if Chinese writings, and especially botanical works, are to be rightly understood.

In order, that Western science may profit by a study of Chinese botanical works, it is necessary not only to understand the Chinese writing, but also to recognize the plants there described. This leads us to a new difficulty. If the plants in question are not generally known, it is for the most part impossible to recognize them from the vague description of the Chinese botanists. Sometimes the good drawings in the Chi-wu-ming &c. permit us at least to determine the order to which the plant belongs. But the only exact method of identifying Chinese names of plants, with their scientific names, is to obtain the plants in natura and to determine them. This is, however, not possible in all cases. As Mr. Sampson rightly observes in his article on Palms (Notes and Queries III p. 131) the carpenter has a (popular) name for each kind of wood he uses, and the woodsman one for each kind of tree he fells; but the names are generally different, and neither the carpenter nor the woodsman is able to identify both, the tree and the wood. In the same manner the Chinese apothecaries know nothing about the origin of the pharmaceutical preparations, they sell in their shops. The medical plants reach the apothecary shops for the most part cut in little pieces or pulverised. It is very difficult to find out the man who collects them, and, in addition to this a great part of the Chinese medical plants grow in Ssŭ-chuan or other provinces hardly visited by Europeans. It is impossible, I believe, to find out, now a days a Chinese. who knows all the plants described in the Pên-ts'ao or at least a great part of them. You cannot even find a gardener, who is acquainted with the all ornamental flowers cultivated in China; each gardener knows only the few plants or trees he cultivates in his garden. But it can be said, that the names of plants, which occur in the Pên-ts'ao, are employed up to the present time in China and well-known by the specialists for the respective plants.

Our botanists, who collect plants in foreign countries do not trouble themselves generally about the indigenous names of the plants and their practical application, and they take no notice of the cultivated plants. Most of the systematic explorers endeavour only to discover new species or to create new genera in order to introduce their name into the science or to call the newly discovered plants after the name of a friend. But in this opinion our botanical authority in China, Dr. Hance, can not be included. Few savants can be found, who embrace all branches of botanical science like this remarkable botanist.

In my opinion it would be more practical, in designating newly discovered plants, to preserve, if possible, the indigenous names, as has been done for instance with Magnolia Yü-lan, Paeonia Moutan, instead of giving them the names of savants or other persons, which often are dissonant or difficult to pronounce. Can anything more ridicu lons be imagined than such names of plants as for instance Turczaninowin, Heineckiana, Müllera, Schultzia, Lehmannia, &c.*

There is a good number of useful cultivated plants in China, which until now are known only by general names or by their Chinese names. How trifling is our knowledge about the numerous Chinese textile plants, which figure in the reports on trade under the name Hemp. At least the articles on this subject by M. Rondot † and Mr. Macgowan (Chinese Repository XVIII, and Chinese and Japan Repository 1863) give very meagre accounts. Only the plant Ch'u-ma (苧麻), Boehmeria nivea, which yields the grass cloth, 夏布 Sia-pu (summer cloth) has been carefully examined.

* The celebrated naturalist Agassiz is right, in complaining (v. the description of his travels on the Amazon river). "Il est pitoyable d'avoir dépouillé ces arbres (palms) des noms harmonieux qu'ils doivent aux Indiens, pour les enregistrer dans les annales de la science sous les noms obscurs de princes que la flatterie seule pouvait vouloir sauver de l'oubli. l'Inaja est devenu Maximiliana,—le Jara un Leopoldinia,—le Pupunha un Guilielma &c.

† I would here observe, that Rondot in adducing a statement of Abel—who says that Sida tiliaefolia, cultivated near Peking, is here called Shêng-ma—seeks to refute Abel, and proves, that the description of the Shêng-ma (升麻) in the Pên-ts'ao (XIII 29) does not agree with Sida. The last fact is true, but nevertheless Rondot is wrong. The book's name of Sida (Abutilon) tiliaefolia is 檾麻 King-ma or 蒪麻 Siang-ma (P. XV 53 Ch. V. XIV). At Peking where it is largely cultivated, especially on the banks of the rivers and canals, it is called 麻果 Ma-kuo (hemp with fruits,) and also Shêng-ma. But in this case the sound Shêng relates to the character 繩 meaning rope. The fibres here are exclusively used for making ropes. This Malvacea attains, in damp places, a height of 10 feet and more, and the large leaves are often 1½ feet in diameter.

There is no people on the globe, which cultivate such a great variety of vegetables and leguminous plants as the Chinese. But the products of Chinese gardens are as little examined by botanists as the textile plants, and we learn from our numerous works on China and its products only, that the Chinese cultivate red beans, black beans, broad beans, ensiform beans or great millet, small millet, black millet &c.

Notwithstanding the works of some French savants, which treat of Chinese dyeing materials, (Rondot, le vert de Chine 1858—Stan. Julien and Champion, industries de l'empire Chinois 1869) we know very little about the plants, which yield these dye stuffs and are for the most part obliged to quote the vague statements of Loureiro from the last century.

Chinese joiners and carpenters use extensively some very precious woods, obtained in China, namely the 楠木 Nan-mu, the 紫檀 Tsü-t'an, the 花黎木 Hua-li-mu, the 紅木 Hung-mu. All these trees are described in the Pên-ts'ao as growing in Southern China, the Hung-mu (red wood) in Yün-nan, the Nan-mu in Ssŭ-chuan, the Hua-li-mu in Hainan, Annam. Father Cibot asserts (Grosier, la Chine II 279) that the tree, which furnishes the valuable Nan-mu is a kind of cedar. But the Pên-ts'ao says, that the leaves resemble an ox-ear. As regards the Hua-li-mu, Mr. Taintor in his interesting accounts on Hainan (Geographical sketch of Hainan, v. Reports on trade in China 1867) says: "several varieties of ornamental woods are obtained, the most abundant of which is the Hua-li, a hard, dark, handsomely veined wood, which is very neatly turned into a variety of articles." Nevertheless all these trees still do not seem to have been determined by botanists. Perhaps specimens of them may be included in the herbariums of our botanists, but they are not identified with the Chinese names.

It occurs often, that the Chinese in different provinces, have different names for the same plant, which must occasion much confusion. But in such a case the Chinese always know besides the local name of the plant, the book name also of the Pên-ts'ao, which they consider as the foundation of botanical knowledge. Li-shi-chên gives also a great number of synonyms of each plant. According to the Pên-ts'ao 大麻 Ta-ma (great Hemp) relates to Cannabis sativa. But in Peking the people understand by this name the *Ricinus communis,*\* and call the *common Hemp* 小 麻 *Siao-ma* (little Hemp).

According to Bridgman's Chrestomathy, Carica papaya, the *Papaw* tree in Canton is called 木 瓜 *Mu-kua* (wood melon). But in Chinese books, as also in Peking, Mu-kua denotes *Cydonia,* the Quince.

For the first knowledge of Chinese botany and Chinese plants we are indebted to the Jesuits, who since the end of the 16th century have preached the gospel in China and studied the country and its people. As a curiosity I will cite a small botanical pamphlet (75 pages) by *Pater Boym,* who lived in China from 1643-59. This first essay in this department, issued in the year 1656 in Vienna, bears the pretentious name "*Flora Sinensis*" but contains only the description of 20 interesting plants and some animals, and 23 drawings with the Chinese characters. This little work is very rare. All accounts of Chinese natural science furnished by the Jesuits (namely by Father *Martini* in his *Atlas Sinensis* 1656) are collected in the admirable work of *Du Halde, Description de la Chine* 1735. There have been described a good number of Chinese plants, animals, minerals, for the most part translations from Chinese books, and also represented by rude drawings.

A work similar to that, drawn up by Du Halde, was published in the year 1818 by *Grosier, Description générale de la Chine* in 7 volumes. Nearly 3 volumes treat of Chinese natural science, 660 pages of them are dedicated to botany. The work of Grosier is most entirely compiled from the *Mémoires concernant les Chinois* and other works of the Jesuits, in the 18th century. Although the articles on Chinese plants of the ancient Jesuits bear no scientific character, they however contain many interesting accounts, either drawn from Chinese authors or the results of their own observation. Grosier has also included in his book a great part of *Loureiro's Flora Cochin Chinensis* published in the year 1790. Loureiro, a Portuguese missionary, described therein a great number of plants of Cochin China and Southern China, joining to the scientific names also the local Chinese names. Notwithstanding the great renown

of this work, being the only Flora of these countries extant, Loureiro, seems not to have possessed the necessary botanical knowledge, for it has been often impossible for modern botanists, to recognize from Loureiro's description the plants determined by him.

Eighteen years after Grosier's work appeared, another compilation on China was published (1836) by 5 authors, *Murray, Crawfurd, Gordon, Wallace, Burnet,* an historical and descriptive *account on China.* Burnet has elaborated the division which treats on natural science, and collected all notes of travellers and naturalists concerning Chinese plants, animals, &c. Therein are also to be found "*Fragments towards a Flora of China.*"

The first, who studied Chinese books on natural history provided with the necessary knowledge of natural science was the well-known sinologue *Dr. S. W. Williams.* Besides several articles on this subject, published in the *Chinese Repository,* Dr. Williams first tried to identify Chinese names of plants, animals and minerals, found in the Pên-ts'ao Kang-mu, with the European scientific names. The three chapters in *Bridgman's Chrestomathy* 1841, treating of Botany, Zoology and Mineralogy are compiled by this remarkable Sinologue.

In the year 1850, *Dr. Tatarinov,* physician of the Russian ecclesiastical mission in Peking, during 10 years, published a list of drugs obtained from the Chinese apothecary shops. Tatarinov, well versed in Chinese, gathered all medicinal plants growing near Peking.\* The plants and drugs collected by him have been examined and determined by special savants in St. Petersburg. This is the origin of *Tatarinov's Catalogus medicamentorum sinensium.* Some of the drugs are described in Gauger's Repertorium f. Pharmacie ẅ. pract. Chemic in Russland 1848. Heft. 12. But a good number of the drugs in the catalogus has not been recognized; and Tatarinov has often made use of Loureiro's diagnosis, which merits but little confidence.

A small, but very valuable work, which treats also of Chinese drugs and medicinal plants, is *Dr. Hanbury's Notes on Chinese Materia medica* 1862.

I would finally mention a small treatise, which likewise endeavours to identify Chinese names of plants with the scientific ones, entitled: *Noms indigènes d'un choix de plantes du Japon et de la Chine* par *J. Hoffmann* et *H. Schultes* 1853. M. Hoffmann says in the preface, that the Floras of Japan and China are very similar, and concludes, that, if the same characters to designate

\* The book name for Ricinus, known only in apothecary shops, is 蓖 麻 *Pi-ma* (P. XVIIa. 32.) The Pên-tsao ranges it under the poisonous plants. It is known, that the seeds, if eaten are very poisonous, whilst the oil extracted from them, the common Castor-oil, is an innoxious purgative. Some assert, that the Chinese use the Castor-oil as food, which loses its purgative action by boiling. As far as I know, the Castor-oil in Peking is only used for lamps and in medical practice. Li-shi-chên explains the character Pi by the resemblance of the seeds with an insect he calls 牛 蝨 (oxen louse.) It cannot be decided from Chinese books whether or not the Ricinus is indigenous in China. The plant is not mentioned before the Tang 618-907. The character Pi is not found in the ancient dictionary Shuo-wên (100 A. D).

\* The hills to the West of Peking are famed for their riches in medicinal herbs, but very many Chinese drugs come also from Ssŭ-ch'uan, Hunan, and Shantung.

plants, occur in Japanese and Chinese books, they denote the same plant. But it is an error. It is true, the Japanese have borrowed from the Chinese their characters for names of plants. These Chinese characters in Japanese botanical writings have the same value as the Latin names of our botanists. There is generally also concordance between Japanese and Chinese plants. But as there are many Japanese plants, which do not occur in China, the Chinese characters for plants are often used in Japan to designate similar plants, or quite different ones. For instance: The character 楓 *Fêng* denotes in China the *Liquidambar formosana*, according to Hoffmann and Sch. it is *Acer trifidum*. 榲桲 *Wên-po* is in China a species of *Crataegus*, much used in sweet-meats in Peking, but *Cydonia vulgaris* in Japan. The 山樝 *Shan-cha*, Crataegus pinnatifida in China H. and Sch. refer to *C. Cuneata*. The name 海石榴 *Hai-shi-liu* for *Camellia japonica*, according to H. and Sch., is, I think, not used in China. The Chinese call the Camellia like the Tea shrub *Ch'a-shu* (v. s.) and they recognized earlier than our botanists (Benth·m and Hooker, genera plant*) that the Cam·llia and the Thea relate to the same genus.—H. and Sch. call the *Aesculus turbinata* (the same as Aesc. chinensis) 七葉樹 *Ts'i-ye-shu*, (seven leaved tree), but as I have stated above this tree is known in Northern China under the name of *Po-lo-shu*. The name Ts'i-ye-shu does not occur in Chinese books. The 紫草 *Tsŭ-ts'ao* of Chinese books is the *Tournefortia Arguzina*, the roots are used for dyeing in red, in Northern China. H. and Sch. state that this name refers to *Lithospermum erythrorhizon*.—H. and Sch. in their list of plants enumerate a good number of plants, which grow only in Japan and therefore cannot have Chinese names. It is, I believe, not proved, that *Illicium religiosum*, the sacred plant of the Japanese, occurs in China (Lindley l. c. mentions it only as a Japanese species) and the name 莽草 *Mang-ts'ao*, which H. and Sch. attribute to Ill. religiostun seems to denote an entirely different plant in Chinese books. See the drawing in the Ch. W. *XXIV*.

Morrison in his Dictionary gives often also scientific names of Chinese plants, but generally they are wrongly adduced. Pru-

dence is therefore necessary in the use of all the above mentioned statements and we cannot "bona fide" adopt the determination of names of Chinese plants by our authors.

The Chinese in their geographical statements generally enumerate plants, beasts and other products of the countries described. These accounts are often very important in enabling us to recognize, what country is meant. Our sinologues, from whom we cannot of course expect a knowledge of natural history, fall often into errors in quoting such wrong determinations of Chinese names of plants.

M. Stan. Julien in his translation of the travels of Wang-yen te to the Oigours (981-983), Mélanges de Géographie Asiatique p. 91, renders the name of a tree 胡桐 *Hu-t'ung*, which occurs in this narrative, by *Volkameria japonica* and 苦參 *Ku-shên* by *Colutea arborea*. I do not know from whence M. Stan. Julien has drawn this information. It can hardly be assumed, that Volkameria japonica grows in the Mongolian desert. The tree Hu-t'ung is said after rain to exude a kind of gum. It is also described in the Pên-ts'ao *XXIV* 64, and represented in the Ch. W. *XXXV*. It is likewise very doubtful whether Ku-shên is Colutea. Loureiro calls *Robinia amara* by this name.

Many errors of this kind are also to be found in a work published in the year 1869 by *M. Stan. Julien* and *P. Champion* under the name, *Industries de l'Empire Chinois*. But these mistakes are however to be ascribed not to the great sinologue, but only to his collaborator, who made his studies in China. I may be allowed, to point out some of these misstatements. M. Champion informs us, that the *Olive-tree* (Oliva europaea) thrives in China (p. 120.) But our olives are not to be found here. The fruit, which bears this name in China is produced from *Canarium pimela* and *C. album*, trees of Southern China. The Chinese name is 橄欖 *Kan-lan* (*P. XXXIb Ch. W. XXXI.*)* The 皂莢 *Tsao-kie*, (black pod, on account of the large black pods) is not *Mimosa fera*, as Champion states, but *Gleditchia sinensis*, (*P. XXXVb* 4. Ch. W. XXXIII,) The 鹽膚子 *Yen-fu-tsŭ* is called by Champion, *Nux . Gallae tinctoriae* (*P. 95*). Mr. Champion meant here probably

---

* Thea olim a Camellia characteribus fallacibus distincta, nuper limitibus certioribus definita, nempe staminibus interioribus liberis numero petalis aequalibus nec duplo pluribus, nobis potius pro sectione habenda, nam genus in integrum servatum magis naturale videtur.

* But in China the *Olea Fragrans* is much cultivated for its little fragrant blossoms, which appear in autumn. The common name is 桂花 *Kui-hua* (cinnamom-flower.) A good drawing can be found in the Ch. W. XXXIII 巖桂.

the 五倍子 *Wu-pei-tsŭ* or Chinese gall-nuts' furnished by a shrub, *Rhus semialata*, called *Yen-fu-tsŭ* by the Chinese (P. XXXII 20. Ch. W. XXXV.) In the same work, there is further described the 地黃 *Ti-huang* (ground yellow) p. 90, a Chinese medicinal plant, used also for dyeing in yellow. Champion calls this plant *Rhemnesia sinensis*. But such a name, I think, does not exist in botanical nomenclature. The same name occurs also in Rondot's work, Notice sur le Vert de Chine 1858. I should say, this is a misprint in Rondot's treatise, which Champion introduced into his own. The Ti-huang of the Chinese is the *Rehmannia sinensis (glutinosa)* of our botanist.—The *Vernicia montana* of Champion is probably the *Elaeococca verrucosa* of botanists, the seeds of which yield the poisonous oil called 桐油 T'ung-yu. Cf. Blakiston's. Five month's on the Yang-tse 1862. M. Champion might have avoided these and other errors, if he had taken the trouble of consulting a generally known and highly useful work, Dr. S. W. William's Chinese Commercial Guide 1863, or Dr. Hanbury's materia medica and other English works. But M. Champion preferred to take information out of French works, written in the last century, as the Mémoires concernant les Chinois, Loureiro's Flora Cochin Chinensis, &c.

## CHINESE ACCOUNTS OF PALMS.

In order to complete my notes on Chinese botanical works and to illustrate my critique of them, I will give some specimens of Chinese descriptions of plants chiefly from the Pên-ts'ao, and I shall choose for this purpose the Chinese accounts of Palm trees, a theme I have already treated briefly in the Vol. III of Notes and Queries (Les Palmiers de la Chine), but which I intend now to present in a more complete form.

I would observe at the outset, that although Palms of several kinds are indigenous in China and now very popular trees among the Chinese, and of great importance, affording many articles necessary to Chinese life and comfort, Palm trees are not however mentioned in the Chinese Cardinal Classics. Neither in the Rh-ya ner in the Shu-king, the Chou-li, or in the Shi-king, which celebrates in song all the renowned plants of the ancient Chinese, can be found any allusion to these splendid trees. The Materia medica of Emperor Sheu-nung makes no mention of any Palm. This is easily understood. The Chinese classics date from the dawn of Chinese civilization, which developed itself in a temperate climate on the fertile soil between two of the largest rivers of Asia, in the *Chinese Mesopotamia*. It was only at the time of Emperor *Shi-huang-ti*, 246-209 B. C., that the Chinese dominions spread to the South of the Yang-tse-kiang and the Chinese made the conquest of the Southern provinces Kuang-tung and Kuang-si, where Palms, the typical trees of the tropics, begin to appear. There is however a Palm in China, the geographical distribution of which reaches to the North as far as the Yang-tse-kiang. This is the Chamaerops Fortuni (xcelsa), and this Palm is mentioned in the *Shan-hai-king* or "Hill and River Classic" (v. s.). It seems therefore to have been known by the Chinese in remote times. The earliest description of Palms by Chinese authors occurs in the *Nan-fang-t'sao-mu-chuang* (4th century), namely of the Cocoa-nut, the Areca Betel, the Caryota and others, and these descriptions are repeated in all botanical works of later time.

## 1. 椰子 *Ye tsŭ.*

*(Cocoa-nut Palm, Cocos-nucifera.)*

(P. XXXI 20. Ch. W. XXXI.)

釋名 *Shi-ming* (Explanation of names).

A synonym for the Ye-tsŭ is 越王頭 *Yüe-wang-t'ou* (head of the ruler of Yüe). According to the Nan-fang-t'sao & c. (v. s.) there is a tradition, that the ruler of 林邑 *Lin-yi* had a quarrel with the ruler of 越 Yüe.* The former sent a man to kill the ruler of Yüe. He found him drunk, killed him and hung his head on a tree. The head became metamorphosed into a Cocoa-nut, with two eyes on the shell.† This is the origin of the name Yüe-wang-t'ou. The Cocoa-nut contains a liquid like wine (the Cocoa-nut milk), and as the Southern people called their rulers by the title 爺 *Ye* denoting "master," they changed also the name of the Cocoa-nut into a name of similar sound, written 椰. Another Chinese Synonym for the Cocoa-nut is 胥餘 *Sü-yü*, a name employed by Ssŭ-ma-siang-ju (2d century B. C.) in his poem *Sang-lin-fu*. Other authors wrote 胥耶 *Sü-ye.*

集解 *Tsi-kie.* (Description of the tree). Ma-chi (an author of the 10th century) says: The Ye-tsŭ grows in 安南 Au-nan (Annam). The tree resembles the Tsung-

* *Lin-yi* was in ancient times a kingdom in India beyond the Ganges, (v. i.) whilst *Yüe* or *Nan-yüe* corresponded with the modern Tonking and Southern China.

† What is commonly called Cocoa-nut is the hard shelled seed of the Cocoa-nut fruit and bears at the base three unequal depressions.

lü (Chamaerops and other Palms, see below). The seed contains a liquid of inebriating properties. *Su-sung* (a writer of the 11th century) states: The Ye-tsü grows in all departments of 嶺南 *Ling-nan*. (Ling-nan, to the South of the Mei-ling mountains, at the time of the T'ang dynasty 618-907, comprised the modern provinces of Kuang-tung and Kuang-si). The *Kuang-chi* (Sung dynasty 960-1280) says: The tree resembles the Kuang-lang, (Caryota sp. v. i.) has no branches, is several 丈 *Chang* high (a Chang=10 feet),* the leaves are like a bundle at its summit. The fruit 實 are as large as a Melon, hanging down between the foliage. The fruit is surrounded by a coarse rind like horse's hair. Within this rind a very hard nut (壳殼) is found, of a roundish and somewhat oblong shape. Within the nut there is a white pulp like pork's grease half an inch thick and more, of a taste like walnuts. This pulp envelopes 4-5 合 *Ko* (about half a bottle) of a liquid like milk, of a cooling and inebriating nature. From the shell different domestic utensils can be made. The white pulp yields sugar. The *Kiao-chou-chi* (description of Southern China) states: The Ye-tsü resembles the Hai-tsung (Ocean Palm v. i.). The fruit is of the size of a large cup and surrounded by a coarse rind like the Ta-fu-tsü (Areca Catechu). In the interior of the fruit is a potable liquor, which does not inebriate. The tree grows in the province of Yün-nan. *Tsung-shi* (an author of the Sung dynasty, 960-1280) repeats the above statements and adds that from the shell wine cups are made. If wine poured into such a cup, contains poison, it will effervesce or the vessel will burst. Nowadays people varnish the inside of Cocoa-nut cups, but then the cups lose their efficacy. *Li-shi-chên* (the author of the Pên-ts'ao) states: The Ye-tsü is the largest of fruits. In planting the Cocoa-nut tree a quantity of salt must be placed near the roots, then the tree will grow high and produce large fruits.

* Mr. Sampson (Notes and Queries III p. 148) quotes a Chinese author, who says, that the Cocoa-nut trees are so high, that men cannot get at the fruit; but they are gathered by the 多羅之人 *To-lo-chi-jen*, who climb the trees for the purpose. Mr. Sampson is inclined to suppose that by the To-lo-jen monkeys are meant. It is true, that in some countries (namely in Sumatra) monkeys are dressed to gather Cocoa-nuts, but in this case men are to be understood. At the time of the Yüan dynasty a wild tribe in the modern Kuang-si and Cochin China was called 多羅蠻 *To-lo-man* (Man=Southern Barbarians.) Cf. Pauthier's Marco Polo p. 431.

It attains a circuit of 3-4 fathoms, a height of 50-60 feet. The tree resembles the *Kuang-lang* (Caryota, v. i.) the Pin-lang (Areca Catechu). It is branchless; the leaves are united at the summit, 4-5 feet long, erect, and point to the heavens. They resemble the *Tsung-lü* (v. i.) and the *Feng-wei-tsiao* (Cycas, v. i.). In the second month bunches of flowers appear between the leaves, 2-3 feet long and as large as 4-5 斗 *Tou* (a Chinese measure of corn). In the same manner subsequently the fruits are arranged in bunches, hanging down from the tree; the largest are of the size of a Watermelon, 7-8 inches long, 4-5 wide. In the sixth or seventh month they ripen. A coarse rind surrounds the fruit. Within is a roundish nut of a dark colour and of a thick, very hard shell. The nut contains a white pulp like snow, of an agreeable sweet taste, like milk. This pulp encloses an empty space, which is filled up by several *Ko* (v. s.) of a liquid. In boring the fructiferous twig a clear fine liquid like wine flows forth. But afterwards it becomes muddy and spoils. The shell of the nut is bright, striated and veined. By slitting it transversely large domestic vessels can be made, whilst by a lengthwise splitting large and small spoons are produced. The History of the T'ang states, that foreigners make wine from the flowers of the Ye-tsü.

These descriptions of the Cocoa-nut given in the Pên-ts'ao are very correct, as every-one will know, who has seen this beautiful and useful Palm. The husk of the fruit yields the fibre, from which the well known *Coir* (derived from the Indian name *Coya* or *Kairu*) is procured, extensively employed in Southern countries in the manufacture of cordage, for matting &c. It is also generally known, that the hard shell is made into various kinds of domestic utensils. Mr. Sampson (l. c. p. 148), states, that in Kiung-chou, the capital city of Hai-nan, great varieties of tea-pots, basins &c., are made from the shells, some simply plain and polished, others more or less highly ornamented with carved figures and of various colours; these are the particular articles of virtu of Kiung-chou-fu. As regards the antipoisonous virtue of these utensils, as mentioned by Chinese authors, this superstition exists also in Ceylon. Mr. Sampson quotes from Yules Cathay II p. 362 the following: "John de Marignolli, early in the fourteenth century, in describing Adam's garden in Ceylon, says of the Nargil (Cocoa-nut): they also make from the shell spoons, which are antidotes to poison.—Li-shi-chên describes also correctly the obtaining of palm-wine from the Cocoa-tree. What is called palm-wine or *Toddy* (this is the

Malayan name, the Indian one is *Sura)* is procured by boring the twigs or by incising the peduncles of the flowers or the unripe Cocoa-fruits. But Toddy can also be made from the sap of other palms, especially the Palmyra palm, (see below). When fermented this palm sap is intoxicating and the best Arrack is distilled from it. By boiling and evaporating it "Jaggery" or sugar is obtained. Some of the Chinese authors seem to confound the Cocoa-nut milk with the palm-wine. As far as I know the milk, an agreeable cooling drink, is not used in the preparation of spirituous beverages.

It is known, that the Cocoa-nut palm is extensively cultivated throughout the tropics of both the old and the new world. Its native country seems to be India and especially Southern India. The Northern limit of its geographical distribution reaches in British India as far as the tropic, but here it grows only on the Western shore, the Eastern shore of British India, and the interior being almost destitute of Cocoa-nut palms. The damp and warm Delta of the Ganges again produces forests of Cocos nucifera, but the tree also does not exceed the tropical limit. In India beyond the Ganges the Northern limit of it extends as far as the 25° of latitude (Cf. Hamilton account of Assam (1798) I p. 243.) As regards China it is known from European sources, that the Cocoa-nut grows abundantly in the island *Hai-nan*, namely on the Eastern coast (Cf. Taintor's Geographical Sketch of Hainan 1868) and forms an article of export trade. On the opposite coast of the mainland, in the Department of *Lui-chou-fu*, the tree also is found. Mr. Sampson states: (l. c. p. 148): "the most northerly spot in which I have seen it flourishing in this part of the world is on the island of Now-chow latitude 20°50.—The Pên-ts'ao asserts, that it grows also in the province of Yün-nan, and in all the departments of Kuang-si and Kuang-tung. But this seems to be an erroneous statement. The great Geography *Yi-t'ung-chi* quotes only the following places as producing Cocoa-nuts: *Kiung-chou-fu* (Hai-nan)— *Yü-lin-chou* (Kuang-si)— *Tai-wan* ( Formosa.)—The *Kuang-si-tung-chi* mentions the Cocoa-nut as a product of *Chên-an-fu* (Kuang-si).

The Cocoa-nut is rich in names. Its Sanscrit name is "*narikela*" (meaning juicy, Cf. Amarakocha, Vocabulary Sanscrit, tradition par Deslongchamps I. p. 115) and has spread to the Persians, Arabians and Greeks, the Persian and Arabian name being "*nar-gil.*" Kosmas Indicopleustes (6th century) calls it "Aggyellion (Cf. Thévenot, Relat. d. voyages curieux 1666 Volume I.) The name

*nyor* used in the Archipelago (Crawfurd, Indian Archipelago I p. 379,) seems to be also of Sanscrit origin. But the Chinese name "Ye" has nothing in common with Sanscrit, and we must be contented with the etymology given in the Pên-ts'ao.—Marco Polo describes the Cocoa-nut, with which he was acquainted in Sumatra (close of the 13th century) under the name of "noci d'India." Cf. Pauthier's Marco Polo p. 573: "Ilz ont moult grant quantité de noix d'Inde moult grosses qui sont bonnes à mangier freshes." The name "Cocos" now the common one among Europeans seems to date from the time Magelhan circumnavigated the globe 1519-22. Pigafetto, the companion of Magelhan, found these fruits first on the Ladrone islands, where they were called "Cocos." (Cf. Sprengel, Pigafetta's Welt-reise 1784.) Bontius (Historia natural Indiae oriental 1631, p. 45) calls the Cocoa-nut "nux indica, a Lusitanis Coquo dicta."

At the end of the description of the Cocoa-nut in the Pên-ts'ao mention is made of three other trees, which the author ranges under the same head.

The 青田核 *Ts'ing-t'ien-ho* (green field nut) is said by *Tsai-pao* (an author of the fourth century) to grow in a country called 烏孫 *Wu-sun.** The tree has a great nut, which, if cut down and filled with water, changes the water into wine of a pleasant taste. This beverage however spoils quickly. Some of this wine was obtained by a ruler of 蜀 *Shu* (an ancient name for Ssü-chuan) towards the close of the Han dynasty (first half of the third century). It is difficult to say what tree here is meant, but it seems to have nothing in common with palms.

The two other trees mentioned, the *Shu-t'ou-tsiu* and the *Yen-shu* relate to other palms, and particulary the Palmyra palm, and will be treated under this head.

I have given in the preceding remarks a literal translation from the Pên-ts'ao, as regards the Chinese accounts of the Cocoa-nut, in order to show the Chinese method of editing and compiling scientific works. But, as the numerous repetitions as well as the unsystematic putting together of the statements would be very tedious for the reader, I will

---

* The *Wu-sun* were a nomadic nation, who lived first on the Western frontier of China (modern Kau-su.) But about 170 B. C. they emigrated together with the 大月氏 *Ta-Yüe-chi* (Ti,) Massagetae, to Western Asia. Cf. Ts'ien-han-shu Hist. of the Ant-Han.1 Chap. 96.

in the further translations set in order the various accounts and quote the names of the authors and the time they wrote only, when they have a particular interest.

## 2. 檳榔 *Pin-lang*,
### *Betel-nut. Areca Catechu.*
### P. XXXI. 15, Ch. W. XXXI.

*Shi-ming* (Explanation of names.) The Nan fang ts'ao mu chuang (4th century) explains the name *Pin-lang* by the custom existing among the people of 交 *Kiao* and 廣 *Kuang* (modern Kuang-tung,) of presenting the Betel-nut to a guest. The character 檳 is formed by the characters 木 tree and 賓 *Pin guest;* the character 榔 *lang* includes the character 郎 *lang* meaning "master," a complimentary term. The Chinese author remarks, that the omission of presenting Betel-nut to a guest would be a mark of enmity. But it seems more likely that the name Pin-lang is a corruption of the Malayan name of the Areca-nut "pinang." As the Chinese language is very poor in sounds and almost every sound relates to numerous hieroglyphs of various significations, it is not difficult in transcribing foreign names by Chinese sounds, to find out characters of a suitable meaning.—Another name for the Areca-nut is 賓門 *Pin-mên* (guest's door.) The poet Ssŭ-ma-siang-ju (second century B. C.) calls the Areca-nut 仁頻 *Jen-pin* (Jen-kernel.) Another name 洗瘴丹 *Si-chang-tan* (the red, washing away distemper,) refers to the sanitary virtues attributed to the Areca-nut.

The Betel-nut has different names in almost every part of Asia. The Malayan name is *Pinang.* According to Sir W. Jones (Asiatic Researches IV p. 312) the Sanscrit name is *guvaca.* Synonyms (given also in the Amarakocha I p. 116,) are *ghónt'á, puga, hapura, cramuca.* The vulgar name in Hindostani is *supyari.* In Javanese its name is *jambi,* in Telinga *Areca.* This latter name was brought by the Portuguese to Europe in the 16th century. The scientific species-name of the tree (Catechu) derives from *Cat'h,* the inspissated juice of a Mimosa, which is chewed with thin slices of the *udvega* or Areca-nut. Sir W. Jones observes, that the Areca Catechu should be called A. Guvafa.—The Arabians know the nut by the name *faufel.* Cf. Voyages d'Ibn Batutu (14th century,) traduit par Sanguinotti, II

204.) The name Betel relates properly only to the leaf of Betel-pepper (see below,) which is chewed together with the Areca-nut, but it is falsely used also to designate the latter.

*Botanical description of the Pin-lang.* By joining logically the numerous statements of different Chinese authors at different times about the Pin-lang, as quoted confusedly in the Pên-ts'ao, we have the following very correct description.

The Pin-lang resembles the Ye-tsŭ (cocoanut tree) and the Kuang-lang (Caryota.) The trunk is straight, branchless, articulated like the Bamboo, 50-70 feet in height. From the top proceed large leaves similar to the leaves of the 芭蕉 *Pa-tsiao* (Banana), which agitated by the breeze sweep the heaven like great fans. In the second or third month a 房 *Fang* (literally a house but here meaning the spathe) arises by a swelling between the leaves, from which, after bursting proceeds a panicle 穗 like the panicle of millet, bearing about 100 white fruits, of the size of a peach or a pear. Below are spines, one over another. The fruits 實 are ripe in the fifth month. They are then as large as a hen's egg, and surrounded by a coriaceous rind 皮殼. Within the rind is a white edible flesh (pulp), which however cannot be preserved in a good state for more than several days, as it quickly spoils. But if treated with lime, roasted or dried in smoke it can be preserved for a long time. The nut 核, within the flesh is veined if broken. It is of a bitter and harsh flavour. The *Fu-liu-têng* (Betel-leaf, see below) and lime must be added, then the flavour becomes soft, sweet and agreeable.

The Chinese distinguish from the form of the nut numerous species or varieties. In the Pên-ts'ao the following are enumerated. A large sort, of a flattened form and harsh flavour is called 大腹子 *Ta-fu-tsŭ* (great stomach) (P. XXXI 19) or 雞心檳榔 *Ki-sin-pin-lang* (fowl's heart Pin-lang) or 豬檳榔 *Chu-pin-lang* (pork Pin-lang). This is used as medicine. A small sort bears the name 山檳榔 *Shan-pin-lang* (hill Pin-lang).—The 蒳子 *No-tsŭ* or 檳榔孫 *Pin-lang-sun* (sun= grand child) is similar to the last but the smallest of all sorts. It is good for eating.* Some Chinese authors speak of a roundish, large and a little conical sort. Our botanists distinguish also several species of Areca, which give edible Areca nuts. I find in Lamark's Botany, I 239: *Pinanga callaparia* Rumph., Areca magno fructu, nucleo subrotundo, acuminato.—and *Pinanga nigra*, Rumph. Areca parvo fructu, nucleo oblongo conico, fuscante. Lindley (Treasury of Botany) mentions *Areca Dicksonii* in Malabar, which furnishes a substitute for the true Betelnut to the poorer classes.

The most ancient Chinese work, which mentions the Pin-lang seem to be the *San-fu-huang-tu*, a description of the public buildings in Chang-an (now Si-an-fu in Shen-si), the Chinese capital at the time of Emperor Wu-ti, 140-86 B. C. There it is stated, that when Yüe-nan (see below) was conquered (B. C. 111.) some remarkable Southern plants and trees were brought to the capital and planted in the Imperial garden (*Fu-li-kung*). Among these trees were also more than 100 Pin-lang. Probably at that time the Chinese became first acquainted with this kind of palm.—*Liu-sün* (an author of the T'ang, 618-907) states, that the best Betelnut is brought by vessels to China and that these growing in China are inferior sorts, namely *Ta-fu-tsŭ̈*. The History of the Liang (502-557) mentions 于陁利 *Yü-to-li* as a foreign country, which produces Betelnuts of a superior quality (Liang-shu Chap. 254, Hai-nan-kuo). There it is said, that Yü-to-li lies on an island in the Southern Ocean. The author of the historical geography Hai-kuo-t'u-chi may be right in assuming, that this realm was in Sumatra. The History of the T'ang (Description of the barbarous regions of the South, Chap. 258ª) names the following as countries, in which the Betelnut is chiefly produced; 環王國 *Huan-wang-kuo*, 哥羅 *K'o-lo*, 眞臘 *Chên-la*, 瀑賦伽廬 *Po-hui-kia-lu*.—The San-fu-huang-tu (first century B. C.) calls 南越 *Nan-yüe* a betelnut growing country.—In the Nan-fang-ts'ao-mu-ch'uang (4th century) it is said, that the Betelnut grows in 林邑 *Lin-yi* and 交趾 *Kiao-chi*. According to other authors it is found also in 扶南

*Fu-nan*.* *Su-kung* (an author of the Tang dynasty) states, that the Pin-lang grows 交州 *Kiao-chou*, in 愛州 *Ai-chou* and in 崑崙 *K'un-lun*. The above mentioned countries refer all to India beyond the Ganges and the Malayan Archipelago. Our botanists agree in the view, that the islands in the Malayan Archipelago (the India aquosa) and especially Sumatra are the native country of Areca Catechu, for it is only on these islands, and the Philippines, that the palm can be found in a wild state. The export of Betel-nuts from Sumatra is enormous. The Betel-nut palm grows also plentifully on the adjacent coasts of the mainland, but its geographical distribution is more limited, than that of the Cocoa-nut. In British India Areca Catechu grows only cultivated and hardly exceeds the tropical limit. To the East from the Malayan Archipelago the growth of the Areca-palm soon ceases.

---

* As these names of countries often occur in Chinese botanical works, I may be allowed to make here a few remarks on those Chinese geographical names, which relate almost all to places in India beyond the Ganges.

In ancient times, up to the time of the Han dynasty (3rd century B. C.) the little known countries to the South of China, namely the Southern borders of the present China, and Tonking, were called by the vague name 南越 *Nan-Yüe* (Southern boundary.) Some Chinese historiographers report that in the year 2350 B. C. an Embassy was sent from 越裳 Yüe-chang to the Emperor Yao. Another Embassy proceeded from this country to the Chinese Court about 1100 B. C. The envoys are said to have brought as presents white pheasants and to have been sent back with a South-pointing chariot. This country Yüe-chang is also identified with Tonking, Cochin China by some Chinese authors (Cf. Li-tai-ti-li-chi VIII 83b. and Hai-kuo-tu-chi.) Others say, that it lay more to the South. (Cf. Pauthier's Relations politiques &c. p. 5 and Dr. Legge's Shu-king. Part II p. p. 533-7.)

交州 *Kiao-chou* comprised at the time of the Han dynasty the modern provinces Kuang-tung, Kuang-si &c. (Cf. Klaproth's tableaux historiques, map No. 7); in later times only a part of Kuang-si and the Northern part of Tonking (v. map. No. 11.) According to the Hai-kuo-tu-chi the Kiao-chou of the T'ang dynasty corresponds with Cochin China and Annam. As the Emperor Wu-ti 140-86 B. C. conquered these countries he established here a Chinese province, of which one district was called 日南 *Ji-nan* (meaning to the South of the sun) and corresponds with the modern Tonking, another, the modern Cochin China,= 交趾 *Kiao-chi* (meaning joined toes, for the inhabitants of this country had crosswise toes.) This name seems to have been the origin of the name Cochin China.—Since the year 679 these countries were called 安南 *An-nan* by the Chinese. The sounds An-nan render the modern name Annam.

愛州 Ai-chou belonged, according to the Geographical Dictionary Li-tai-ti-li-chi (VII i.), at the time of the T'ang to the modern Annam,

---

* Purefoy Cursory states: (Asiat. Journ. 1827 XXII p. 143 Remarks on Cochin China.) In Cochin China are 3 kinds of Betel-nut, a red, a white, and a small kind, which is much exported to China.

As regards the growth of the Betel-nut in China I will quote the following from the Pên-ts'ao and other Chinese works. The most ancient description of this palm, in the Nan-fang-ts'ao &c., (4th century) does not say, that it thrives in China proper. The writers of the T'ang and Sung (7-12 century) state that it grows in all departments of 嶺外 Ling-wai, (beyond the Mei-ling mountains, the modern Kuang-tung and Kuang-si.) The geography of the Sung dynasty notices the Pin-lang as a tribute of Kiung chou (Hai-nan.) The island of Hainan produces Betel-nuts extensively up to the present time. Mr. Sampson (l. c. p. 133) states that Ling-shui, on the South coast, produces the best. According to Mr. Taintor

The name 林邑 Lin-yi ( Land of forests ), known to the Chinese since the 3rd century A. D., is described in the History of the Liang (6th century) Chap. 54. It is said there that Lin-yi lies on the borders of Ji-nan (v. s.), and was called Yüe-chang (v. s.) in ancient times. The capital is distant 120 li (3 li—1 English mile) from the Sea and 400 li from the boundary of Ji-nan. To the South Lin-yi is bordered by water (Sea?) Klaproth identifies on his maps Lin-yi with Siam. Ritter (Asien III. 977) with Cochin China. The Wên-sien-tong-kao (14th century) states that 環王國 Huang-wang-kuo and 占城 Chên-ch'êng are other names for Lin-yi.

扶南 Fu-nan lies, according to the same work (Liang-shu) 7000 li to the South of Ji-nan (Tonking) on a bay 灣中, which stretches to the West of the sea. From Lin-yi it lies to the South-west, 3000 li distant. The capital is situated 500 li from the sea. There is a large river to the N. W. of it, 10 li broad, which flows to the East in the sea.—Abel Rémusat (Nouv. Mél. asiat. I 77) states, that by Fu-nan Tonking is meant. The Hai-kuo-tu-chi identifies Fu-nan with 暹羅 Sien-lo, or Siam. Although it is impossible to determine with certainty the position of Fu-nan from the vague Chinese description, there can however be no doubt, that it was a place in India beyond the Ganges. I venture moreover to observe that perhaps Fu-nan lay on the banks of the Mekong. Crawfurd states (Cf Ritter l. c. III. p. 914) that the province Sadek in Cambogia is called Fo-nan in the Cochin Chinese language.—Since the time of the Sui dynasty 589-618, Cambodia was known to the Chinese by the name 真臘 Chên-la. In the History of the Sui it is said, that Chên-la was formerly dependent upon Fu-nan. Its position is given as to the South East from Lin-yi. The sea forms its Southern boundary.

崑崙 K'un-lun is the ancient name of a range of mountains in Central Asia, but the Chinese use these characters also to designate the island Pulu Condore near Cambodia.

哥羅 K'o-lo or 哥羅富沙 K'o-lo-fu-sha lies according to the T'ang history (Chap. 258a) to the S. East of the 盤盤 P'an P'an, but about this country it is there said, that it lies on the sea, to the S. West of Lin-yi, from which it is separated by a little sea. From Kiao-chou it can be reached by ship in 40 days. Therefore it can be assumed, that by K'o-lo an island near Malacca or in the Malayan Archipelago is meant.

(l. c. p. 14) the Areca palm flourishes in the Eastern and Southern parts of the island. The land on which it is grown is subject to the payment of a land tax.—The great Geography of the Chinese Empire, Yi-tung-chi, states further, that the Betel-nut thrives in Tai-wan (Formosa,) in the department of King-yüan-fu in the province of Kuang-si, (according to the Kuang-si-t'ung-chi also in Chên-nang-fu in the same province.) in Yüan-kiang-chou in Yün-nan. The special Geography of Yün-nan notices also Lin-an-fu, Kuang-nan-fu, as Betel-nut countries.—The French explorers of the Mékong (Revue des deux mondes 1870 p. 340) have seen the Areca Catechu near Yüan-kiang, in Yün-nan (23½° latitude): "La ville de Yuen-kiang, assise au bord du fleuve (Sonkoï) était entourée de champs de riz a demi coupés, de bois d'aréquires, de champs de canne à sucre &c."

In the tropical countries, where the Areca palm thrives, there is to be met everywhere another plant closely connected with the Betel-nut, however not by botanical alliance, but only by the combined use made of both plants by the people of these regions. The Betel chewing nations can hardly imagine the Areca-nut without the leaf of Betel-pepper, which has given its name even to the nut. The Betel or Areca-nut is prepared for chewing by cutting it into narrow pieces, which are rolled up with a little lime, obtained from oyster-shells, in leaves of the Betel-pepper. This pellet is chewed and has formed for a long time an indispensable dietetic requisite and healthy regulator of all classes of men in Southern Asia. It is known, that by Betel chewing the saliva is tinged red. It stains also the teeth and is said to produce intoxication in the beginning. The Betel-pepper, Chavica Betel (another species Chavica Siriboa is used for the same purpose,) is a twining plant with large oval acuminate shining leaves, and flowers in long spikes. It belongs to the order of Piperaceae and is widely cultivated in tropical and intertropical Asia, so that its native country now can not be fixed.

The common Chinese name for the Betel-leaf is 蒌 Lou or 蒟 Kü. According to Bridgman's Chrestomathy the second character is pronounced Lau in the Canton dialect. In the Pên-ts'ao the Betel-pepper is described ( XIVa 46 ) under the name of 蒟醬 Kü-tsiang. Li-shi-chên explains that it regulates the digestion. Therefore the first character includes the character 蒟 meaning "strong," the second means "Soya."

Another name is 土蓽茇 *Tu-pi-pa* (indigenous Pi-pa or Long pepper,) another 扶惡土蓽藤 *Fu-ô-tu-lou-t'êng* (the character T'êng means twining shrub, the other characters express probably a foreign name.) Another Synonym is 扶櫚 *Fu-liu* or 扶櫚藤 *Fu-liu-t'êng*, about the origin of which, Li-shi-chên declares, he knows nothing. These names do not resemble any name given to this by other Asiatic people. The Sanscrit name of the Betel-plant is, according to the Amerakocha (l. c. p. 105) *nagaralli*, the name of the leaf is *tambulavalli*. The Arabians call it *tenbol* (cf. Ibn Batuta l. c. II 204.)—Büshing (Asien II 764-783,) states that at Malabar the Betel-leaf is called *Wettilei;* the Indo-persian name is *pan.* Bontius states regarding Betel (l. c. p. 90.) "Folia ista quae Malaii *Sirii* vel *Sirii-bou* vocant, Javani *Betel.*"

Among the Chinese works quoted in the Pên-ts'ao about the Betel-pepper the Nan-fang-ts'ao &c., (4th century) is the most ancient. The description of the plant, given by the Chinese authors of various times is the following. The plant climbs like the cucumber, the leaf is large, thick, shining and of a pungent, aromatic taste. The fruit resembles that of the mulberry, but it is of a long shape, several inches long. These leaves are eaten together with the Pin-lang (Areca-nut) and calcined oyster shells. It has the property of expelling distemper and to make one forget sorrow. In Ssŭ-chuan an inebriating beverage is made from the 蓽葉 *Lou-ye* (leaf of the Betel-pepper.

As regards the native countries of the Betel-pepper the Chinese authors notice *Kiāo-chou, Ai-chou* (Annam see the foot note above.) An author of the 11th century states, that the plant grows in *Kui-chou* (Ssŭ-ch'uan) in *Ling-nan* (Kuang-tung, Kuang-si.) According to other authors it is also found in *Yün-nan.* The Pên-ts'ao, states further, that there are several kinds of Betel-pepper. The Ch. W. gives (XXV p. 45) a tolerably good drawing of the Chavica Betel under the name of *Kü-tsiang* and represents (XXV 49) the *Lou-ye* as a different climbing plant with large heart-shaped leaves. I am not able to state, whether the true Betel (Chavica Betel) thrives in Southern China as the Catholic missionaries assert (Grosier la Chine II 225.) Bentham in his Flora Honkongensis mentions several species of Chavica, namely *Chavica sarmentosa* (formerly determined as Chavica Betel by Seemann,) found also in Java, Bor-

neo, New Guinea, and adds, that, besides the shape of the leaves, this is at once known from the Chavica Betel by its remarkably short spikes. Is this the Betel used by the Chinese for Betel chewing?

Crawfurd (History of the Indian Archipelago) is of opinion, that the use of Betel as a masticatory, originated in the Sunda islands, and has spread from thence to the Asiatic continent. The antiquity of the use of Betel among the nations of Southern Asia can not be determined with certainty. The Persian historiographer *Ferishta* states, that about 600 A. D. in *Kanyakubja,* the capital of the Duab (Northern India) there were 30,000 shops, which sold the Betel-leaf (pan) Cf. Ritter's Asia IV I. p. 859. *Ibn Batuta,* an Arabian traveller, who visited Hindostan in the 14th century, describes the process of Betel chewing there (l. c. II. 204). He calls the Betel-leaf *Tenbol.* The names Areca and Betel, generally used by European writers to designate the nut and the leaf, were introduced by Pigafetto, the companion of Magelhan, the circumnavigator of the globe, 1519-22. Pigafetto states (Sprengel l. c. IV. 53.); "the inhabitants of the Messana island (Philippines) cut a pearlike fruit, which they call *Areca* into four pieces and roll them up with a Laurel-like leaf called Bettre. This is chewed by them &c."

The Betel is now-a-days much used as a masticatory in the Southern provinces of China. Even at Peking the Areca-nut is well known and sold everywhere in the streets. But as the Betel-leaf used for chewing must be in a fresh state, the Chinese in the Northern provinces restrict themselves to eating the Betelnut alone. The practice of Betel chewing was not known by the Chinese in ancient times, at least it is not mentioned by the writers of the Han dynasties. But in the History of the Post-Han (25-221. A. D.) mention is made of very distant islands, inhabited by the 黑齒 *Hei-chi*(blacktoothed men). This seems to be an allusion to the nations, which chew Betel (Cf. my article Fu-sang, Chinese Recorder III. p. 114).

The *Long pepper,* Chavica Roxburghii, is also mentioned in Chinese books. The Pên-ts'ao describes it very correctly (XIV. 44) under the name of 蓽撥 *Pi-pa.* There it is said, that Pi-pa is a foreign name. A writer of the 8th century states, that the name of the plant in the Kingdom of 摩伽陀 *Mo-kia-to* (the ancient Maghada in the present province of Bengal) is 蓽撥梨 *Pi-pa-li,* whilst in 佛蒜 *Fo-lin* it is called 阿梨訶陀 *A-li-ho-to.* *Pi-pa-li* is the Sanscrit name of Long pepper; another Sanscrit name is *Chavica* (Cf. Amarakocha l. c. I. 99.100). Bontius (l. c. p. 182) says : " Bengalenses Pimpilim nuncupant, quod nos, auctoritate graecorum Piper longum." To what language A-li-hó-to must be referred, I am not able to say. Fo-lin designates, as is known, the Greek Empire. The plant is described by the Chinese authors as follows: The pi-pa belongs to the Betel genus. The leaves are shining thick and circular and resemble the Betel-leaf, the stem is like a tendon, the root is black and hard. The flowers are white, appear in the 3rd month, the fruit is long, like a little finger, of a greenish, blackish colour. In the 9th month it is gathered and dried in the sun. Its taste is like *Hu-tsiao* (Black pepper). The 胡 人 *Hu-jen* (Western Barbarians) like to mix it with their food. The plant occurs also in 波 斯 *Po-ssŭ* (Persia) and in 嶺 南 *Ling-nan* (provinces of Kuang-tung and Kuang-si) where it grows in Bamboo-forests. This description suits quite well with the *Chavica Roxburghii*, a climbing plant with oval shining leaves, which is largely distributed in India. Long pepper consists of the spixes of flowers, which, while yet immature, are gathered and dried in the sun. There spikes and the roots are employed as medicine by the natives. The Jesuits confirm the statement of the Pên-ts'ao, that Long pepper is produced also in Southern China (Cf. Grosier, la Chine, II. 525).

The Fathers Boym and Martini (17th cent.) assert further, that the *common Pepper* (Piper nigrum) is cultivated in the Chinese province of Yün-nan (Grosier, l. c. II. 519). The same is stated in the Pên-ts'ao, where Black pepper is described under the name of 胡椒 *Hu-tsiao* (XXXII. 9). There it is said, that in *Mo-kia-to* (Maghada) it is called 昧 履 支 *Mei-lü-chi*. This name can be referred either to *Maricha*, the Sanscrit name of Black pepper, or to *Mirch*, its name in Hindostani. I cannot find among the numerous Sanscrit synonyms of Black pepper, as given in the Amarakocha (l. c. I. p. 2 11.) a name, resembling the Greek péperi, from which originate all names of Pepper in the modern European languages. Hippocrates (5th century B. C.) states, that the Greeks received this product and the name peperi from the Persians. But the Persian name of Black pepper is *Filfil*. In my opinion the name peperi was wrongly applied in ancient times to Black pepper, for it seems to be derived from the Sanscrit *Pi-pa-li*, which relates to Long-pepper.

## 3. 無 漏 子 *Wu-lou-tsŭ.*

### *The Date Palm. Phoenix dactylifera.*

### P. XXXI 22.

*Shi-ming.* Explanation of names. The Date Palm bears according to the Pên-ts'ao a great number of synonyms, of which Li-shi-chên gives the following explanations. The tree is called 波 斯 棗 *Po-ssŭ-tsao* (Jujube from Persia) for it grows in Persia. (As regards Po-ssŭ-kuo I beg to refer to my article Chin. anc. geograph. names, Notes and Queries IV). The fruit is called 苦 魯 麻 *Ku-lu-ma.* (By these sounds the Persian name of the Date, being " Khurma " is rendered as correctly as possible by Chinese characters). The names 千 年 棗 *Ts'ien-nien-tsao* (thousand years Jujube) and 萬 歲 棗 *Wan-sui-tsao* (ten thousand years Jujube) are an allusion to its vigorous growth and long-lived character. The names 蕃 棗 *Fan-tsao* (foreign Jujube), 海 棗 *Hai-tsao* (Ocean Jujube) and 海 櫻 *Hai-tsung* (Ocean Palm) relate to its foreign origin and to the resemblance of the fruit to the Jujube (Zizyphus vulgaris). It is further called 金 果 *Kin-kuo** (golden fruit) in allusion to its utility and high value.

*Description of the tree.* Li-shi-chên states, after a writer of the Ming (1368-1644), that near *Chêng-tu* (the capital of the province of Ssŭ-ch'uan) there are six Kin-kuo trees, of an aged appearance, planted at the time of the Han dynasty (about our era). They are 50-60 feet in height, three fathoms in circumference. The stem is erect like an arrow, without lateral branches. The leaves are like a phoenix tail. The bark resembles dragon's scales, the fruit is like a Jujube, but larger. Its foreign name is *Ku-lu-ma* (v. s.) The author adds. that the fruit becomes edible only (he speaks apparently of the Ssŭ-ch'uan fruit) after a treatment with honey and other complicated processes. —This description suits quite well with the Date-palm. It is known that the stem is marked with scars, indicating the places from which the leaves have fallen away in proportion as the tree has grown in height, and at the top new leaves unfolded. These

* I must observe, that now a days the fruit of Salisburia adiantifolia bears also the name of *Kin-kuo* (Jinko in Japanese.)

scales render it very easy to climb the tree. It is also true, that the Date resembles much the Jujube and for this reason also the Europeans call the large Chinese Jujubes, Chinese Dates. That the fruit of a Date-palm growing in Ssŭ-ch'uan cannot be edible is also easily understood; for it is a fact, that the fruits of the Date-palm ripen only in a rainless climate. Chêng-tu lies under the 30th degree of latitude, in a climate, which permits the thriving of a palm tree, planted in favourable conditions.

The *Kuang-hün-fang-pu* (Chap. 79 p. 14) quotes two works of the 12th century, which mention, also some rare trees, called *Hai-tsung* (Ocean Palm) at Chêng-tu-fu. There it is further stated, that once an attempt was made to transplant them to 金陵 *Kin-ling* (an ancient name for Nan-king.) But they could not grow in the climate of Nan-king and had to be brought back to Chêng-tu. These trees were carefully treated there and protected against injury by a wall.

The *Hai-tsao* (Ocean Jujube,) which is said by Li-shi-chên to be identical with the Date-palm is described in the repeatedly quoted Nan-fang-ts'ao &c., (4th century) as follows: An erect tree without lateral branches. The branch-like leaves on the top of the tree diverge in every direction. The tree bears fruit only once in five years. The fruit is as great as a cup and resembles, a Jujube. The Kernel is not pointed at the ends, as the Jujube. It is rolled up from the two sides. The Hai-ts'ao is sweet and well tasted, superior to the Imperial Jujube in the Capital. In the year 285 A. D. Lin-yi (a kingdom to the South of China) (see above) offered to the Emperor Wu-ti (Tsin dynasty) 100 trees of the Hai-ts'ao. The prince *Li-sha* told the Emperor, that in his travels by sea he saw fruits of this tree, which were, without exaggeration, as large as a Melon (!)

Under the name of *Po-ssŭ-tsao* (Persian Jujube) or *Wu-lou-tsŭ* the Date is first described in Chinese works of the 8th century. These authors state, that the tree is found in Persia, where it bears the name 窟莽 *Kü-mang* (probably a distorted transcription of khurma) It is said to resemble other Palms, as regards the stem and the leaves, which do not fall in winter and are in shape like the leaves of the 土藤 *Tu-t'êng* (probably a Rattan.) It flourishes in the second month; the blossoms resemble the Banana blossoms. It opens grad-

ually (the spathe,) and some ten bunches spring from them. Each cluster (朵) has 20-30 fruits. The fruit is 2 inches in length, at first of a yellowish white colour and like the fruit of the 楝 *Lien* (Melia Azedarach.) It ripens in the 6th-7th month and then becomes dark, resembling the fruit of the 青棗 *T'sing-tsao* (dark Jujube) from Northern China, but the flesh (pulp) is crumbling. It is of a sweet taste like sugar and has the colour of the 沙糖 Sha-t'ang (brown impure sugar.) The kernel differs from the kernel of the Jujube by the absence of the pointed ends (the kernel of the Jujube is very pointed.) It is rolled up from the sides ( 隻卷.) The Po-ssŭ-tsao is brought to China in vessels by merchants from those countries, where it grows.

The description here given of the Date-palm and particularly of the fruit and the kernel is very correct. There can therefore be no doubt, that the Po-ssŭ-tsao is the Date. But it is clear, that many of the synonymns, as given in the Pên-ts'ao, relate often to other Palms, which is easily understood, for the Date-palm is not indigenous in Eastern Asia, and, although some Chinese writers assert, that it was planted here in ancient times,—now-a-days, I think, the Date-palm occurs nowhere in China. The Pên-ts'ao gives a good drawing of it, but the *Ch. W. (XXXII)* represents under the name of Wu-lou-tsŭ a palm with fan-shaped leaves. Nevertheless it is certain, that at the time of the T'ang dynasty (618-907) the Date-palm and its fruit were well known in China. Embassies were often sent from the Persians and the Arabians to the Chinese court and even Chinese envoys and travellers visited the Date growing countries. (See my article: Chinese Ancient Geographical Names in Notes and Queries No. 4.) During the Yüan dynasty (1286-1368) and the Ming dynasty 1368-1644 likewise relations existed between China and those countries of Western Asia.—Mr. Sampson quotes a Chinese author, who states, that the Dates (Ts'ien-nien-ts'ao) come from 忽魯謨斯 *Hu-lu-mu-ssŭ*. As I have proved in Notes and Queries (l. c. p. 53) the country here meant is *Ormuz* in the Persian Gulf. Ritter (Asien VI p. 724) is of opinion, that the name Ormuz is derived from the Persian word "khurma," (Date,) for the Date-palm grows plentifully on the shores of the Persian Gulf.

As regards the geographical distribution of the Date-palm it is a representative of the subtropical countries of Western Asia and the Southern littoral of the Mediterranean. It is confined to the more dry zones, where vehement rains do not exist. Therefore the Date grows plentifully in Northern Africa, Arabia, Southern Persia, Beloochistan, and the North Eastern corner of British India (Punjaub, Lahore, Moultan.) But here is the Eastern limit of its distribution. To the South it can be found as far as Bombay, but here the fruits do not ripen.—In Persia it is only the Southern provinces, which produce dates, namely the littoral of the Persian Gulf and Kirman. The most Northern spot in Persia, where the Date is cultivated, is the oasis *Tubbes* in the great Salt-desert (about 34° latitude.) But at Isphahan, which has a more Southern position, I have not seen Date-palms. There is however in Mazanderan, (Ashref) near the shore of the Caspian sea a splendid Date-palm, which was planted by a Persian Shah, some centuries ago.—Bagdad (33° latitude) produces good Dates.

Mr. Sampson (l. c. p. 172) mentions a species of Chinese Phoenix (or Datepalm) in the following terms: "A species of Phoenix grows wild in Hongkong and generally near the sandy shores and slopes of the hills along the sea coast; it is often nearly stemless, but when suffered to grow to full development, has a cylindrical caudex of from two to six feet in height; this is referred doubtfully in the Flora Hongkongensis to Ph. acaulis, Roxb., but Dr. Hance (Seemann's Journal of Botany, Vol. VII p. 15.) shews it to be *Ph. farinifera* Roxb." Mr. Sampson adds, that he is not aware that this plant has at all attracted the attention of the Chinese. But in Dr. Hance's adversaria in stirpes Asiae orient. p. 48, I find a description of this palm: "Species Hongkongensis generis Phoenicis, quae videtur diversa ab omnibus, quas descripsit Griffith. Propinqua autem videtur Ph. sylvestri, Roxb. Crescit gregarie in petrosis aridioribusque collium lateribus." Dr. Hance adds, that the fructiferous spadices of this palm are sold in Macao under the name of "Areca de mato" (i. e. Areca sylvestris,) and that the Chinese eat the farinaceous fruits, which however are very adstringent. Phoenix farinifera is common all over India and grows almost together with Phoenix sylvestris (the wild Date.) The stem of Ph. farinifera yields in India a meal, a substitute for the true Sago. It is used especially in time of famine (Ritter, Asien IV$^a$ p. 862.) Lamark mentions Ph. farinifera as growing also in Cochin China.

4. 棕櫚 *Tsung-lü* Chamaerops, and 蒲葵 *P'u-k'ui*, Fan Palms.

*P. XXXV, 78. Ch. W. XXXV.*

*Shi-ming.*—Explanation of names. A popular manner of writing the above name is 棕櫚 *Tsung-lü.* Both names are derived from 騌鬣 *Tsung-lü* (horse-hair,) on account of the fibres, like horse-hair, which surround the bark. Another name of the tree is 栟櫚 *Ping-lü.*

*Description of the tree.*—The authors quoted in the Pên-ts'ao about the Tsung-lü (for the most part writers of the 10th century) and Li-shi-chên himself following description of it:

The Tsung-lü is a tree in height, about the same thickness perfectly straight and branchless. leaves, which grow all from the tree, spread out from thence like a fan, in every direction. They resemble in shape a carriage wheel and do not fall in the winter. At first, when the leaves begin to unfold they resemble the 白及.* The leafstalks are three-cornered. An author of the 8th century says, that in Ling-nan (Southern China, see above.) there are several trees, the leaves of which resemble the Tsung-lü, namely the *Ye-tsu* (Cocos nucifera,) the *Pin-lang* (betel-nut,) the Kuang-lang (Caryota spec. see below,) the *To-lo* (Borassus, see below,) the 冬葉 *Tsung-ye* and the 虎散 *Hu-san.* †

* The 白及 *Po-ki* refers to an Amomaceae, according to Tatarinow (Catal. med. chu.) Indeed the drawing of the Po-ki in the Ch. W. VIII 12 seems to represent a species of Alpinia.

† I am not able to state what trees are meant by Tung-ye and Hu-san.—Of the *Tung-ye* (winter leaf,) the following short account is given in the Nan-fang-ts'ao ku, (4th century.) The Tung-ye, called also 薑葉 *Kiang-ye* (Ginger leaf,) or 芭蕉 *Pao-tsu* is used in Southern countries. The climate there is very hot and everything spoils quickly. This can be prevented by wrapping them in the leaves of the Tung-ye. Things can be preserved in this manner for a long time.

The 虎散 *Hu-san,* called also 古散 *Ku-san* is described in a few words in the Pên-ts'ao at the end of the article Kuang-lang (XXXI p. 24.) There it is said, that from this tree canes are made. This is perhaps *Rhapis flabelliformis,* a palm native of Southern China, with fan-shaped leaves. Lindley (Treasury of Botany) states about this palm, that it is said to yield the walking canes known as Ground-Rattans. Mr. Sampson, however, says (l. c. p. 172) that

Below the place, where the leaves proceed, there is a fibrous integument, formed by several strata of entangled fibres. When one circuit has ceased growing, it forms a joint on the stem. The trunk is of a reddish brown colour; the wood is fibrous and veined. It can be used for stamps and for manufacturing domestic utensils. In the 3rd month, from amidst the leaves at the top of the trunk, there issue several yellow bunches, formed of young flowerbuds, in appearance like fishroe; therefore they are called 椶 魚 Tsung-yü (yü=fish;) another name is 椶 筍 Tsung-sun (sun=Bamboo sprouts.) These bunches gradually expand and form a large panicle (花 穗) of light yellow flowers. In the 8th or 9th month the fruits are formed. They are abundant and crowded together in large racemes. The fruit is about the size of a bean and of a yellowish colour as long as unripe, but when ripe black and very hard. The Chinese consider the Tsung-lü as a tree of great utility (大 爲 時 利.) Besides the above mentioned use of the wood, the fibres are woven into various articles of domestic use, clothing, hats, cushions, mats to sleep on &c. Ropes are also made from the fibres, which ... ive injury by many years im... ... water. The Chinese authors ... ... the fibres must be removed from ... or three times a year, for they ... the growth of it. By omitting to do ... ... But the Kiang-kün-... ...) cannot be cut off ... too frequently or the tree ... gured. The same work quotes an ... of the 11th century who states, that in 閩 She (the modern province of Ssü-... ) the Tsung-sun (the flowerbuds of the tree, v. s.,) gathered in the first or second month, are used as food, especially by Buddhist priests, who prepare them by boiling like Bamboo sprouts, &c.—The Pên-ts'ao states finally, that in Southern China, there can be distinguished two kinds of the Tsung-lü tree, the one bears a fibrous integument, used for making ropes, the other is smaller, without fibres; its leaves can be used for brooms. Some authors were of opinion,

that this smaller kind of Tsung-lü and the 王 蔧 Wang-sui are the same. But Li-shi-chên proves, that Wang-sui is another plant, identical with the 地 膚 Ti-fu.*

The Tsung-lü seems to be the only Palm known to the Chinese in the most ancient times, at least the character 椶 Tsung occurs in the Shan-hai-king or "Hill and River Classic," which the Chinese attribute to the Emperor Yü (2200 B. C.) It is there said that at the 石 萃 Shi-tsui hill and at the 天 帝 Tien-ti hills a great many Tsung trees grow. The ancient Chinese Botany Nan-fang-tsao &c. (4th cent) mentions the tree as 栟 櫚 Ping-lü.

As regards the geographical distribution of the tree in China, according to Chinese sources, the ancient Chinese works, quoted in the Pên-ts'ao, mention it as growing in Ling-nan (Southern China beyond the Mei-ling mountains) and Ssü-ch'uan. It is further said there, that it was planted also in Kiang-nan (the modern Kiangsi and Fukien,) but it did not grow easily. In the Wu-lü-ti-li-chi (T'ang dynasty 618-907) it is stated, that on the hills near 臨 沅 縣 Lin-yüen-sien there is an abundance of Ping-lü trees. [Up to the T'ang dynasty the modern Wu-lin-sien (Chang-tê-fu in the province of Hu-nan) was called Lin-yüan-sien. Cf. Yi-tung-chi.]

In the great geography of the Empire, Yi-tung-chi, and in the special description of the single provinces I find the following departments and districts mentioned as producing Tsung-lü trees.

*Province of Che-kiang:*—Hang-chou-fu—Shao-sing-fu (Shan-yin-sien)—Tai-chou-fu (Ning-hai-sien.)—Kü-chou-fu (Chang-shan-sien.)—Yen-chou-fu (in all districts.)

*Province of An-hui:*—Chi-chou-fu.—Liu-an-chou.

*Province of Hu-nan:*—The Tsung-lü tree is generally mentioned in the Hu nan-tung-chi.

*Province of Kiang-si:*—Nan-an-fu.

*Provinces of Kui-chou* and *Yün-nan* (generally mentioned.)

*Province of Kuang-si:*—Kui-lin-fu.

---

Rhapis flabelliformis is known at Canton by the popular names 椶 竹 Tsung-chu (Palm bamboo) or Chu-tsung (Bamboo palm,) and that it is a tree of no importance or celebrity. Rhapis flabelliformis is described in Bentham's Fl ra hongkongensis. The synonym Rhapis kwanwortsik Herm. Wendl. quoted therein seems to be derived from a Chinese name of the plant in the Southern dialect.

* Ti-fu or 掃 帚 草 Sao-chou-tsao (Broom plant,) P. XVI 44. Ch. W. XI. 31, is the Kochia (Chenopodium) Scoparia. This pretty shrub grows everywhere at Peking and is much cultivated also in gardens where it takes the shape of a dense bushy globe.

Mr. Sampson may be right in assuming, that the name 欏 *Tsung* has become to some extent a sort of generic term in popular language for Palms in general. As the character *Tsung* is said in the Pên-ts'ao to be derived from another character meaning horse-hairs, I think, the Chinese apply it to all Palms, which yield horse-hair-like fibres, namely Chamaerops, Livistonia, Rhapis, Caryota. Mr. Sampson is also correct in stating, that the names *Tsung-lü* and the synonym *Ping-lü* relate not to one only, but to several Palms. The Chinese themselves distinguish several species of Tsung-lü. But it seems to me that, nowadays at least, these names relate more especially to the Hemp-palm, Chamaerops Fortuni. This is proved by the geographical distribution pointed out for the Tsung-lü or Ping-lü in the above quoted geographical works. The Chamaerops is the most Northern genus of Palms. *Ch. humilis*, the African and European representative of this genus, extends as far as Nice to the North (43½° of lat.) Ch. Palmetto grows in Northern America, namely in Georgia. A third species is found in Japan and was described by Thunberg (1784) as *Chamaerops Excelsa* but mentioned much earlier by Kaempfer (1712.) Some 20 years ago Fortune detected in Northern China the Chinese Hemp-palm, named in the system *Chamaerops Fortuni*. But some botanists believe, that this is only a variety of Chamaerops excelsa. Fortune mentions repeatedly this beautiful Palm in his writings on China and gives also a good drawing of it. Fortune, in visiting the Tea countries in China, met with the Hemp-palm in the Northern provinces, namely in Chekiang, on the island of Chusan and in An-hui. He states, that near Ningpo the hills are covered with it. Fortune says further, that in the countries, where this tree is found, the Chinese agricultural labourers use the coarse brown fibre, obtained from the hairlike bracts(!) for making ropes, hats, bed-bottoms, and also the garment called So-e 蓑衣, known as, "raincloaks" by Europeans, worn in wet weather and protecting perfectly from the rain. These accounts given by Fortune agree perfectly with the above description of the Tsung-lü from Chinese sources. The Chinese say, that the Tsung-lü is a tree of 10 to 20 feet in height. According to Fortune the Hemp-palm grows to about 12–20 feet in height. All species of Chamaerops are more or less dwarfish palms. There can be no doubt, that the Palm tree Tsung-lü or Ping-lü mentioned by the Chinese authors as growing in the Northern provinces, namely Chekiang, An-hui, Hunan, Kiangsi, can be other than the

Palm described by Fortune. But it is possible, that in Southern China, where the Chamaerops is not indigenous, (Fortune,) the name Tsung-lü is applied to other Fan-palms, which give fibres. Mr. Sampson states, that in the province of Kwang-tung, under the name of Tsung-lü, two kinds of Fan-palms are cultivated, a coarse leaved= *Livistonia chinensis* R. Br., and a fine-leaved. The latter is commonly, though whether correctly or not, Mr. Sampson is not prepared to say, said to be the Chamaerops excelsa, Thbg. Mr. Sampson says further, that the fine leaved species (Chamaerops,) when distinguished from the coarse (Livistonia) is termed 蒲葵 *P'u-k'ui*, the latter character giving its name to the fans, *K'ui-shan*, into which its leaves are made. I am not aware, whether in China fans are made from the leaves of Chamaerops; Fortune does not mention it, and the Pên-ts'ao says also nothing about the manufacture of fans from the leaves of this Palm. But the Pên-ts'ao as well as the Kuang-kün-fang-pu describe *P'u-k'ui* as a peculiar Palm, growing only in Southern China, from the leaves of which fans are made, and distinguish it clearly from the Tsung-lü.

At the end of the article Tsung-lü the Pên-ts'ao states: But the 蒲葵 *P'u-k'ui* is a different palm. Li-shi-chên does not agree with the ancient Dictionary *Shuo-wen* (100 A. D.) which considers the *P'u-k'ui* identical with Tsung-lü. Li-shi-chên gives the short description of the P'u-k'ui, consisting of 13 characters, as found in the Nan fang ts'ao mu ch'uang (4th century). The P'u-k'ui resembles the Ping-lü (Chamaerops) but the leaves are finer. Fans can be made from them. The P'u-k'ui grows in 龍川 *Lung-ch'uan*, (Province of Kuang-tung, Hui-chou-fu).

I think the P'u-k'ui must be another Palm than the Chamaerops. The leaf fans made from the leaves of the P'u-k'ui palm, and known in commerce under the name of 葵扇 *K'ui-shan*, (Cf. Dr. Williams' Commercial Guide p. 119) form an important article of trade. Mr. Sampson states: "The leaf fan is said to have been first introduced into use among the élite of the Northern provinces, during the Tsin dynasty (A. D. 265-419) when the barbarian people of the South are stated to have attached great value to the products of the *Tsung* tree; the wind from these fans was supposed to be peculiarly agreeable; and it appears, that at that time these leaves came into special repute, for it is stated, that hats were made from them, which were worn by men of all classes and superseded the turbans formerly in use. In the manufacture of certain kinds of hats they are still employed in Canton. According to the *Kwang-tung Sin-yu*, in the preparation of leaves for fans, the finest are selected, soaked in water for a fortnight, and then redried by fire heat. This process gives them a smooth polish; they are then bordered with silk or rattan fibres and fastened at the junction with the stalk by brass pegs driven through plates of shell; just, indeed, as we find them at the present day."

The Chamaerops Fortuni has been introduced by its discoverer in England and is now also cultivated in France. It is perfectly hardy in the Southern parts of England and grows in the open air in the gardens of Cherbourg, Bordeaux &c. ( Bull. d. l. soc. d'acclim Juillet 1869). In Peking it is much cultivated, but not in the open air, the winter in Peking being very rigorous.

Loureiro describes also the Chamaerops Cochinchinensis, as growing in Cochin China. I am not aware, whether this is a true Chamaerops or rather another Palm.

Our European writers have often mentioned in their works on China the manufacturing of garments, mats, ropes &c. from palmfibres, but their accounts about the origin of these fibres present much confusion. Morrison (Dictionary of the Chinese language) says: the *Tsung* is a tree, of the bark of which the peasants make garments to defend them from the rain. Dr. Williams in his Middle Kingdom, I. 278, states: "The fan leaf palm (Rhapis) is cultivated for its leaves. The wiry fibres of the bracts (!) of the Rhapis are separated into threads and used largely for making ropes, cables, twine, brooms, hats, sandals and even dresses or cloaks for rainy weather. Dr. Williams' Commercial guide p. 86: The most of the coir is made from the bark of the Hemp-palm ( Chamaerops ). The loose bark is stripped off in large sheets from the trunk of the tree, and when steeped in water the fibres separate in short wiry threads of a dark brown colour. It is the material, from which the Chinese make mats, brooms, cordage, raincloaks &c."

Fortune states, that the raincloaks are made from the bracts of Chamaerops.

Finally I find in the Bull. d. l. soc. d'acclim 1862, No. 4, a very curious statement. There it is said: "à Canton il y a une espèce de Chanvre (!) appelée Chamaerops excelsa ou Hemp aloös (!) dont on fait le po lo ma

pu." Dr. Williams ( Commercial Guide ) states, that the Po-lo-ma-pu is made from the fibres of a Corchorus.

I am of opinion, that the textile fibres in question are neither obtained from the bark of palms, nor from the bracts of it. ("Bracts" is a botanical term used for the leaves placed immediately below a calyx and altered from their usual form). But, as the Chinese authors correctly state, the base of the leaf stalks (of several palms) is enveloped by a fibrous integument, the fibres of which are entangled and cross each other. These fibres seem to proceed from the base of the petioles. After the leaves have fallen off, the remains of the leafstalks and the leafsheets separate themselves in fibres and form the above mentioned network. This process I have observed myself on the specimens of Chamaerops, cultivated in Peking, but I am not acquainted, from my own observation, with the mode of obtaining these fibres by the Chinese for the purpose of manufacturing garments, ropes &c.

## 5. 桃椰子 *Kuang-lang-tsŭ.*

*Caryota species.*

P. XXXI 23, Ch. W. XXXI.

*Shi-ming.*—Explanation of names. The name Kuang-lang is said by Li-shi-chên to be derived from 光 *Kuang*=smooth and *Lang*=Betelnut, for the tree resembles in appearance the Areca palm and has a smooth stem (or wood). The wood especially is called 姑椰木 *Ku-lang-mu.* The synonym 麵木 *Mien-mu* (flour-wood) refers to the meal contained in the stem, the synonyms 董櫚 *Tung-tsung* (solid palm) and 鐵木 *Tie-mu* (iron wood*) have refer-

---

* This Ironwood must, however, not be confounded with the *Ironwood* of Loureiro, *Baryxylum rufum*, sinice *Tie-li-mu* (Flora Cochinchinensis). Grosier (la Chine II 236) and Duhalde (la Chine I p. 24) give the following description of it, after Loureiro and other missionaries:—This tree, which grows in several provinces of China, is as high as our oaks. It is remarkable for its wood, which resembles iron as regards the colour as well as its hardness and heaviness. It cannot float in the water. The tree belongs to the Leguminous order, has pinnate leaves, yellow flowers with 5 petales, 10 unequal stamens. The flowers are arranged in racemes, the fruit is a long pod, a little curved, roundish and contains several seeds. The Chinese use the wood as timber in all cases, where great loads must be supported and great resistance is required. It is from the Tie-li-mu, that the anchors of the Chinese ships are made. Dr. Williams in Bridgman's Chrest. p. 441 quotes the Tie-li-mu under the name of iron pear wood among Canton woods. But our botanists consider this tree, described by Loureiro under the name of *Baryxylum* as dubious. In Bentham and Hooker, Genera plant I p. 464 it is said: "Baryxylum Lour. est genus valde dubium. Description auctoris

ence to the durability and strength of the wood.

I find the Kuang-lang first described in the Nan-fang-ts'ao &c., (4th century,) but mention is made of it in the History of the Post–Han 221-263 A. D. The Chinese authors describe this palm as follows:

The trunk is 50-60 feet in height, several fathoms (!) in circumference, upright, without lateral branches. The tree resembles the Pịng-lü (Chamaerops) the Pin-lang (Areca

pluribus notis Cassiam refert, Icon. Rumphii dubie citata est Aſzeliae species. Specimen Loureirianum, errore quodam sub hoc nomine in IIb, Mus. Brit. servatum cum characters nequaquam convenit."— Chinese books give but little information as regards the Tie-li-mu. It is not mentioned either in the Pěn-ts'ao or in the Kuang-kün-fang-pu. Only in the Chi-wu-ming &c. (descriptive part XXI b p. 100, article 欄 木 Lü-mu) I found a short account of the Tie-li-mu, taken from the Nan-yüe-pi-ki, a description of the modern Kuang-tung province. There it is said:

In Kuang-tung there are three kinds of wood used in carpentry, the 紫檀 Tsü-tan, 花梨 Hua-li and 鐵力木 Tie-li-mu.

An author of the 4th century says, that the 紫檀 Tsü-tan comes from Fu-nan (in India beyond the Ganges v. s.) The name Tsü-tan (red tan) is explained by the brown red colour of the wood and the resemblance of the tree to the Tan tree. The character Tan refers according to Hoffmann and Shultes (l. c.) to a Caesalpinia. The drawing of the Tan in the Ch. W. XXXV agrees quite well with this. 檀香 Tan-siang is the Chinese name for Sandalwood. The Tsü-tan used at Peking is of a dark brown colour and very heavy.

花梨木 Hua-li-mu (the meaning of the characters is wood veined like pear-wood) is a common name. The book name of this tree is 欄木 Lü-mu. The wood is a little fragrant, of a brown red colour curiously veined. If the tree is old, the lines are more curved, on young trees they are straight. Delineations can often be found like coins. The tree grows in Annam and also in Hainan, namely in Yai-chou (Southern coast) Ch'ang-hua (North-western coast,) Ling-shui (South-eastern coast.) I have already stated above, that the growth of Hua-li-mu in Hainan is confirmed by European writers. The Hua-li-mu, which is sold at Peking, is a very beautifully veined wood of a brown colour. Cf. also Grosier (l. c. p. 288).

As regards the 鐵力木 Tie-li-mu (wood of the strength of iron) only a few words are dedicated to its description in the Chinese work. There it is said, that it is very durable and hard. The colour of the wood is at first yellow, but becomes after use, black. In the 黎山 Li-shan hills the people use it for the fuel. But when it arrives at the Northern provinces it becomes very dear. Li-shan here refers probably to Hainan, for the aborigines of the island are called 黎 Li. But Li-shan is also a hill in Honan.

Catechu,) the Ye-tsŭ (Cocoa-nut,) the Po-ssŭ-ts'ao (Date-palm.) The wood is hard like Bamboo wood, of a dark brown colour, very durable. It is veined like the Hua-li-mu (see the foot note.) The centre of the trunk is humid and rots quickly. The joiners cut it into little pieces and make chess-boards from it. It is adapted also for shovels and spades. In some places the mariners use spears of Kuang-lang wood. On the summit of the tree, large leaf like branches and luxuriant racemes of greenish flowers proceed. The fruits can be gathered throughout the whole year. They resemble blackish pearls and are produced abundantly. One branch contains not less than 100 fruits and each tree has 100 of such branches, which hang down gracefully. The whole resembles an umbrella. Below the insertion of the leaves, there is a net-work of entangled horse hair-like fibres, resembling the fibres of the Tsung-lü (Chamaerops.) The Kuang-tung people collect and use them for manufacturing tissues. But they must be at first soaked for some time in saltwater in order to become fine. These fibres are also used for ship building. The author adds 不用釘線 "they use neither nails nor threads." Mr. Sampson explains this passage by a quotation from Yule's Cathay: "Menentillus, a Dominican Friar, writing from Southern India in A. D. 1292, says: their ships in these parts are mighty frail and uncouth, with no iron in them, and no caulking." The bark of the Kuang-lang tree is very tenacious and flexible. It serves to make ropes. The Chinese authors, who describe the Kuang-lang, agree in stating, that within the bark of the tree a white flower (according to some authors of a yellowish red colour,) is found, resembling pounded rice. It is said to be very nutritious. The Chinese say, those, who eat the Kuang-lang flour, will not suffer from hunger. In the provinces, where the Kuang-lang grows, corn is there not abundant and therefore people eat the Kuang-lang flour with cow's milk or bake it into cakes. The flour is found several inches beneath the bark. A large tree yields 100 Chinese pounds of it.

Ancient and modern writers agree, that the Kuang-lang grows in the Southern provinces of China. According to the History of the Post–Han (25-221 A. D.) the Kuang-lang tree is found in 句町縣 Kü-ting-sien (now-a-days Lin-an-fu province of Yünnan,) and flour is obtained from its trunk. The Nan-fang-ts'ao &c., (4th century) states Kiao-chi (Cochin China, v. s.) and 九眞

Kiu-chên as its native country. (Kiu-chên was at the time of the Han a district in the modern Annam. Cf. Li-tai-ti-li-chi IV, I.) Another ancient work (Yi-wu-chi,) says that the tree grows in 牂 牁 Yang-ko, (Yang-ko comprised in ancient times parts of the modern provinces of Ssï-chuan, Hu-kuang, Kui-chou, Kuang-si: Cf. Biot. 1. c).— The Chung-nan-chi 14th century,) quoted in the Kuang-kün-fang-pu, says: in the three districts (郡) of 梁 水 Liang-shui, 與 古 Sing-ku and 西 平 Si-ping there grows little corn. But the Kuang-lang which yields flour is found there. These three districts comprised in ancient times the Western part of the modern Kui-chou province and the North Eastern part of Yünnan. (Cf. Li-tai-ti-li-chi IX 4, XII 10, XIII, 2). Su-sung (an author of the 11th century) states, that the Kuang-lang grows in Ling-nan (v. s.) and in all districts of Kuang-tung and Kuang-si, where it is much cultivated in gardens. Li-shi-chên indicates Ssŭchuan, Kuang-tung, Kuang-si, Annam as the native countries of the Kuang-lang.

According to the great geography of the Empire and the special descriptions of the single provinces I find the Kuang-lang mentioned as a product of the following provinces and districts.

Yünnan. Kai-hua-fu—Kuang-si. Nanning-fu, Wu-chou-fu, Ssŭ-chêng-fu, Chên-nang-fu.—Kuang-tung. According to the Kuang-tung-chi there is a hill 60 li to the North of Lien-chou, where a large number of Kuang-lang grow.—Ssŭ-chuan. Sü-chou-fu. The Kuang-lang here is found on the hills 石 門 山 Shi-mên-shan.

I am not acquainted myself with the palm, which the Chinese call Kuang-lang. But Mr. Sampson in his interesting article on palms gives some accounts of it, based upon personal inspection, which I may be allowed to quote here. Mr. Sampson identifies the Kuang-lang with a species of Caryota, whichhe saw growing abundantly in Shui-tung on the West coast, (Mr. S. does not say what West coast he means), and which is planted at Canton in monastic and temple grounds for ornament. Along the bank of the West River it is abundant, and may frequently be seen rearing its graceful head above the other trees of natural woods; on the border of Kuang-si is a magnificent grove formed entirely of these trees. In Canton the Caryota is almost always called Tsung. ( But as has been pointed out in treating of the Fan palm this is in popular language a generic term for Palms, which

yield horse—hair like fibres). The name Kuang-lang is (now at Canton) seldom given to the Caryota tree, but the fruits of it are sold in druggist's shops under the name of Kuang-lang-tsŭ (Tsŭ-seed). The identity of the Caryota with the Kuang-lang of Chinese authors is not quite free from doubt, for the plates of this tree given in the Pên-ts'ao and in the Ch. W. (both represent a palm with fan shaped leaves) do not accord with the Caryota, and the statements of authors, that Sago is made from the pith of the tree, are not verified, as far as Mr. S. can ascertain, by the practice of the Southern Chinese of the present day. But it must be remembered, that the tree intended to be represented grew only in Southern China, in Cochin China and perhaps other adjacent countries, the whole of which territory was, at the time, when the original accounts of the tree were written, loosely classed as the Barbaric states of the Southern ocean; it is highly probable therefore, that the Northern Chinese authors never saw the tree, and only figured it in accordance with imperfect descriptions, filling up the gaps by drafts on their own imagination.—As regards the flour obtained from the pith of the tree, according to the Chinese, there is nothing at all improbable in the statement, that a Caryota can yield a farinaceous product, for another representative of this genus (Caryota urens) in India is known as a Sago yielding Palm. In addition to this the Caryota is the only Palm in Southern China according to Mr. Sampson, to which the Kuang lang can be referred.

Mr. Sampson states: The most important product of the Kuang lang at the present day is the fibrous sheaths or bases of leafstalks; this is the Tsung fibre of native commerce of Canton. It is principally imported from Kuang-si. Mr. S. describes the raw material as follows: they are in the form of an isosceles triangle, about eighteen inches in length and ten inches wide at the base; they are composed of fibres, longer than those of the Cocoanut tree, crossing each other in two directions with considerable regularity; the apex, which represents the lower end of the leafstalk, becomes somewhat ragged, and the base, which represents the downward continuation of the leaf stalk forming a portion of the trunk, is covered with a fine thin cuticle, which however soon wears off. The uses to which these fibres are put are mainfiold; the entire sheaths are employed in covering boxes, securely fastened down by small ropes made of the same material; some of the ropes used in ships, and smaller ropes for all purposes, are twisted from the fibres, and are said to be remarkable for their power of resisting the injurious effects of long immersion in the water. Brooms are also made from them.

## 6. 莎木麪 *So-mu-mien.*

### P. XXXI. 23. Ch. W. XXXV.

This is another tree, resembling the Kuang-lang, which the Chinese authors describe as yielding flour. It is more than 100 feet in height, the leaves proceed from the summit of the tree and spread in two directions like a flying bird. Another author compares the arrangement of the leaves to the 莎衣 *So-e,* or raincloaks (the character so denotes the material, from which raincloaks are made, v. s.) Hence the name *So-mu-mien* (Mu=tree, mien=flour.) The latter character relates to the white or yellowish white flour, obtained from the bark (or the pith) of the tree. This flour is smooth and better than that derived from the Kuang-lang; cakes can be made from it. One tree furnishes about 100 pounds of flour. An author of the 8th century states, that the So-mu-mien grows in Ling-nan (Southern China.) According to the Shu-ki (Annals of Ssǔ-ch'uan, 15th cent.) it is found also in Southern Ssǔ-ch'uan.

The Wu-lu-ti-li-chi (T'ang dynasty) describes a tree 櫅木 *Siang-mu* which resembles the Kuang-lang and which yields a white flour like bruised rice. This tree is said to grow in Kiao-chi (Cochin China, v. s.). Li-shi-chên is of opinion, that this tree and the So-mu-mien are identical. He identifies it also with the tree 都勾 *Tu-kü,* mentioned in the Kiao-chou-ki, as resembling the Kuang-lang and yielding flour.

It is difficult to say, what tree here is meant by the Chinese authors. In Southern Asia there are many trees, the trunk of which yields a granulated form of Starch, known under the name of Sago. The true Sago, sent to Europe is obtained from *Sagus laevis* and *S. Rumphii,* native of the Molucca islands. In Ceylon and Malabar it is obtained from *Corypha umbraculifera,* in Malabar also from *Caryota urens.* All these palms cannot be considered as the Sago-palms, described by Chinese authors, for they occur only in countries distant from China. According to Dr. Williams' Commercial Guide, nowadays the true Sago, brought to China from Singapore is known by the Chinese as 西穀米 *Si-ku-mi* (Western corn rice.) As Dr. Hance states (Notes and Queries III. p. 95:) "no true Sago-palm has hitherto been detected in China, though one of them, *Arenga saccharifera,* occurs in Cochin-China. But there are in China or in the adjacent countries some representatives of the genus *Cycas,* separated by the modern botanist from the true Palms, which furnish Sago. Loureiro, writing of his *C. inermis* (which only attains a height of about 5 feet) states, that it is not used for food in Cochin China, but he adds: Tunkini incolae mihi retulerunt in sua patria fieri panem Sagu sat bonum ex trunco hujus palmae. Thunberg again says (Flora japon p. 280) of *Cycas revoluta* (much cultivated also in China) but which also rarely exceeds the height of a man: medulla autem caudicis supra modum nutriens, imprimis magni aestimatur; asseverant enim, quod tempore belli frustulo parvo vitam diu protrahere possint milites, ideoque in commodo eodem fruatur hostis extraneus, sub capites poena vetitum est palmam e regno japonico educere*.*

In India Sago is obtained also from *Phoenix farinifera,* a dwarf palm, which occurs also in Southern China, as has been above stated. But the Chinese assert, that the Sago-palms, known by them and used for food are of high growth. All the Cycas species, with the exception of C. circinalis, which attains a height of 40 feet, are also of a dwarfish nature. In addition to this, several species of Cycas are known by the Chinese and described in their books under other names, as I will point out subsequently. It is therefore unlikely, that by the name of So-mu-mien, or Siang-mu the Chinese understand a Cycas or Phoenix farinifera. But perhaps the Siang-mu, which is said to thrive in Cochin China, means the *Arenga (Saguerus) Saccharifera.* This Sago-palm is mentioned by Loureiro under the name of *Borassus gomutus* as growing in the forests of Cochin China. (Cf. Lamarck, Botanique).

## 7. 梖多 *Pei-to.*

*Borassus flabelliformis, Palmyra palm.*

(The first character is sometimes written 貝.)

This name is applied by some ancient Chinese authors to the *Sacred Fig* (Ficus religiosa) but more generally it relates to a Palm tree, namely the *Palmyra palm,* Borassus flabelliformis, and *Corypha umbraculifera.*

The *Pei-to* tree is mentioned repeatedly in the 佛國記 *Fo-kuo-ki,* the well known work of the Chinese Buddhist priest 法顯 *Fa-sien,* who visited during the years 399-414 A. D. the countries, where Buddha was worshipped. Fa-sien seems to refer this name always to the *Ficus religiosa.* He saw (l. c. p. 23-24) the Pei-to-shu, beneath which past and future Buddhas attain perfection near 伽邪 *Kia-ye.* This is the ancient *Gaya* in Maghada, where Sakyamuni lived 7 years until he attained to Buddhaship. —Another Buddhist priest 玄奘 *Hüan-tsang,* well known among our savants, who travelled over India in the first half of the 7th century, mentions this sacred tree in the same place and calls it 道 樹 *Tao-shu* (the tree of intelligence) a literal translation of the Sanscrit name "Bodhidruma." Cf. Stan. Julien, Mémoires s. l. contr. occid. II. 376. The same tree, in the shade of which Buddha is said to have spent 7 years of penance, exists still, a splendid Sacred Fig tree; 2 miles S. E. of Gaya in Bahar.

Fa-sien (l. c. p. 29) states further about the Pei-to-shu: The ancient Kings of the 獅 子 國 *Shi-tsŭ-kuo* (lion's kingdom, a literal translation of Singhala, the ancient name of Ceylon) sent a deputation to 中 國 Chung-kuo (Middle kingdom*) for seeds† of the Pei-to-shu (貝 多 樹 子.) The Pei-to-shu were planted near the temple of Buddha. As the (principal) tree attained a height of 200 feet it inclined to the South-East. The King, being anxious that it should not fall down ordered it to be supported by 8 or 9 pillars. The tree shot forth then a branch, which after having grown through one of the pillars, descended and took root in the ground. Fa-sien says, that the tree was 4 圍 *Wei* ‡ in circumference at the time he saw it. The pillars, although curved and cracked also still existed. There seems to be no manner of doubt, that Fa-sien speaks not of a palm tree, but of Ficus religiosa,—although the statement, that a branch of the tree descended and took root in the ground, points more to the Banyan tree, Ficus indica.

---

* By Middle kingdom Fa-sien understood not China, but, as he explains himself (l. c. p. 5) India. He calls China always by the name of its celebrated dynasties 漢 *Han* and 晉 *Tsin.*

† Baal, Travels of Fa-sien p. 152 translated "a slip of the Pei-to-shu."

‡ As the character *Wei* occurs very often in Chinese descriptions of trees and some of our sinologues wrongly understand this word, I will give a short explanation of it. Landresse states in a note to Rémusat's Fo-kuo-ki, p. 344, "4 Wei environ 0,m.0612. Ls-wei équivaut à la moitié d'un *Tsun*, lequel est la dixième partie de la coudée Chinoise, soit 0,m 0106." Hence it would follow that the splendid Pei-to-shu, several centuries old, which Fa-sien saw in Ceylon was of the size of a walking cane. I do not know from what sources Landresse received this information, but I find in the Dictionary of Kang-si the following

五 寸 曰 圍 一 抱 曰 圍 *Wu-ts'un-yŭo-wei-i-pao-yŭo-wei.* Five Ts'un (inches) or tenths of a cubit) are called *Wei*, and also one fathom. Morrison translates in his Dictionary the character *Pao* wrongly by bundle, but its meaning is "to embrace" or the distance between the horizontally extended arms of a man (a fathom). Such contradictory meanings of the same character occur very often in the Chinese language, which, notwithstanding the high position assigned to it by the eminent savant W. v. Humboldt ( Verschiedenheit des menschlichen Sprachbaues), is one of the most imperfect and confused.—But the character *Wei* in Chinese botanical writings denotes always a fathom and not 5 inches.

---

The tradition of Fa-sien, regarding the introduction of the Pei-to tree from India to Ceylon is met with also in the ancient annals of Ceylon (Cf. Sacred and historical books of Ceylon by Upham, 1833, III. 219, a detailed account of the transportation of the branch of the Bogana tree at Anuradhepura.) There is described how a branch of the sacred Bo tree, beneath which Buddha entered "nirvana," was brought with great ceremonies from Maghada, the fatherland of Buddha (Sakyamuni) to Singhala (Ceylon) and planted in the garden Mahomenah near Anuradhepura (288 B. C.) Cf. also Chapman's remarks on the ancient city of Anarajnpura in the Transactions of the Royal Asiat. Soc. Vol. III. p. III. The same tree is still at the present day an object of veneration by Buddhists. Cf. Tennent's Ceylon II. 613.

*Ficus religiosa*, the Peepul tree, the Sacred Fig tree of the Buddhists, one of the giants of the vegetable kingdom, is considered throughout India as a sacred tree. Burmann in his Thesaurus ceylanicus 1737 describes it as Arbor zeylanica religiosa foliis cordatis, integerrimis acuminatis, prepetue mobilibus Boghas, Budughas incolis dicitur. The trembling of the leaves of the tree, like the Aspen tree, is a characteristic of it, often mentioned and poetically interpreted in ancient Buddhist works. Ficus religiosa is called *Bodhi* (meaning intelligence) by Northern Buddhists, or *Chadula* (the tree with tremulous leaves), in Hindustani = *Pipala* (Cf. Amarakocha l. c. I. p. 84). In Chinese Buddhist works the name Bodhi is rendered by the characters 菩 提 *Pu-t'i* and Pipala by 畢 鉢 羅 *Pi-po-lo.* Cf. Kuang-kün-fang-p'u. Chap. 81 p. 7. A fine drawing of it is found in the *Ch. W. XXXVII. p. 27.* Besides these names, Chinese Buddhists call the tree 恩 惟 樹 *Ssŭ-wei-shu* (tree of meditation.) As has been done also often by our botanists in former times, the Ficus religiosa is confounded by some Chinese authors with the *Ficus indica* or *Banyan tree,* * for some authors state, that the roots of the Pu-t'i-shu grow from the branches.

---

* *Ficus indica, the Banyan tree,* is another sacred tree of India, but more especially an object of veneration by the Brahmins. A striking characteristic of it and distinguishing it from Ficus religiosa is, besides the oval lanceolat leaves, that the branches send roots down to the ground, which form new trunks. In this way one tree forms a whole forest. The Banyan tree is found throughout India, in Ceylon, the Archipelago, to the West as far as Arabia. Loureiro mentions it in Cochin China (Ficus indica, ramis latissime expansis radices crassas in terram demittentibus). Neuhoff (Gesandtschaftsreise nach China, 1666), describes and represents the Banyan tree and states, that he saw it growing in China. Ainslie (Materia med. ind. II p. 10-11) asserts, that the Banyan tree is called *Yang-shu*

I suppose, that Fa-sien by the characters[8] *Pei-to* intended to render the name of the Bodhi tree. It was only after the time of Fa-sien, that the characters P'u-t'i for rendering this name came into use.

On the other hand some Chinese authors chose the characters 貝多 *Pei-to* to designate the Palms, or rather the leaves of Palms, which are used in India for writing (Palmyra palm). Chinese writers explain, that Pei-to means leaf (patra) in Sanscrit. But the Palmyra palm bears the Sanscrit name *Ta-la*.

Mr. Sampson (l. c. p. 180) gives the translation of several quotations in the Kuang-kün Fang-p'u, regarding the Pei-to-shu. As I am not able to present to the reader, a more correct translation, I may be allowed to quote Mr. Sampson's words, adding only a few explanations.

In China. Aluslie means probably 榕樹 *Yung-shu.* Under this name, which does not occur in the Pên-ts'ao, the Kuang-kün fang-p'u describes pretty well the Banyan tree, as a large wide branching evergreen tree, with numerous rootlets pendant from the branches, which on reaching the soil penetrate it and form, as it were, new trunks, so that a large tree will have roots in 4 or 5 different places. A single tree will afford a shade of several now in extent. A fine drawing of the Yung tree is found in the Ch. W. XXXVII 10. Mr. Sampson gives much interesting information about the Banyan tree in China. (Notes and Queries III p. 72.)

"The Banyan tree of South China, as the distinguishing name Bastard-banyan, which is often applied to it, imports, is not considered identical with though it is closely allied to the celebrated Banyan tree of India. According to Flora Hongkongenis the Bastard banyan is the *Ficus retusa* L. In Southern China there is scarcely a rural ferry landing on the rivers of Kuang-tung, that is not furnished with one or more, to afford shelter to the passengers as they await the return of the boat; few public buildings are without the tree to adorn and shade the space in front or the court yards behind. There is no doubt, that the Banyan is a native of this part of the world. In China it extends northward as far as the Yang-tsze; it is abundant in the Fu-kien province, and has for that reason given its name to the capital city Foochow, which is poetically termed 榕城 *Yung-ch'êng* or Banyan city; and besides growing abundantly in the more Southern provinces, it forms a prominent feature in the landscape along the rivers south of the Poyang lake.

Mr. Sampson states, that the Yung tree is mentioned only by modern Chinese authors. But he overlooked the fact, that the first book quoted in the Kün-fang-p'u about the Yung tree is the Nan-fang-ts'ao-mu-ch'uang (4th century) and that some of the statements, which Mr. S. translates, are taken from this work.

The Sanscrit name of Ficus indica is *aswattha*. By this name it is always called in the Vedas, Shastras, Puranas and other ancient Indian writings. Kreeshna said:—'The Eternal Being is like the tree aswattha, the roots of which turn towards the heaven, whilst the branches descend to the ground. (Cf. Bhaguat-geeta or Dialogues of Kreeshna and Arjoon, quoted in Ritter's Asien IV 2, p. 665). This points unmistakably to Ficus indica. Some savants, however, consider aswattha as a synonym for the Bodhi tree (Ficus religiosa). Cf. Amarakocha l. c. I p. 84, also Asiatic Researches Vol. IV p. 309. Mr. Eitel (Chinese Buddhism I p. 25) uses also the name aswattha (阿濕

The *Shi-wei-ki* (4th century) states. In 洛陽 *Lo-yang* (the Chinese Capital during the Tsin dynasty 265-420 A. D., to the West of the present Ho-nan-fu) the Yih Tsin bridge leads to the Bôdhi-manda (altar of intelligence, v. Eitel l. c. p. 25) where the Buddhist classics were translated. At this Bôdhi-manda were upwards of ten Brahmin and Indian priests making a new translation of the classics, the originals of which came from abroad and were written on leaves of the *Pei-to* tree; the leaves are one foot and five or six inches in length, and more than five inches broad; in form they are like a 琵琶 *Pi-pa* (guitar) but thicker and larger; they are written on crosswise, and are bound together in books of various sizes.

The *Yu-yang-tsa-tsu*, or Desultory Jottings of Yu-yang (close of the 8th century) reads as follows:

The 根多 *Pei-to* tree comes from Magadha (v. s.); it is sixty or seventy feet in height, and its leaves do not fall in winter. There are three kinds of this tree:

1. 多羅婆力义貝多 *To-lo-p'o-li-ch'a-pei-to.*

2. 多利婆力义貝多 *To-li-p'o-li-ch'a-pei-to.*

喝咂 *A-shi-ho-ta* in Chinese books) as a synonym for the Bodhi tree. It seems indeed, that the two sacred trees of India, Ficus religiosa and F. indica are often confounded by native writers. Other Sanscrit names of Ficus indica are *Vata* and *Nyagrodha* (Cf. Asiatic Researches IV p. 309, also Amarakocha l. c. p. 88). The latter name is rendered in Chinese Buddhist books by 尼拘律 *Ni-kü-lü.* This tree Ni-kü-lü is mentioned also by Fa-sien (p. 24), besides the Pei-to-shu, as a tree, beneath which Buddha sat on a square stone, turned to the East. Perhaps some of the quoted synonyms relate to other species of the Genus Ficus. Sir W. Jones in the Asiatic Researches IV p. 109 enumerates 4 kinds of holy Fig trees in India, distinguished by different Sanscrit names, Ritter in his Asia, IV 2, p. 656-88, gives very valuable accounts of the sacred Fig trees in India.

Besides these trees Buddhist works enumerate some other trees, in connexion with the different Buddhas, namely.

The *Pandarica* (Bignonia spec ?)
The *Patala* tree (in Chinese 波吒釐 *Po-to-li*), the *Trumpet-flower* (Bignonia suaveolens, according to Wilkins.)

The *Sal* tree 娑羅樹 *So-lo-shu* in Chinese). *Shorea robusta.* Sakyamni's (Buddha's) death took place in the shade of Sal trees. Mr. Eitel (l. c. p. 114 commits an error in identifying Shorea robusta with the Teak wood. Teak wood is obtained from *Tectonia grandis.*

*Sirisha* (in Chinese 尸利沙 *Shi-li-sha*; Cf Pên-ts'ao XXXV 3. Article 合歡 *Hô-huan*). *Mimosa Sirisha*, according to Roxburgh.,

### 3. 部婆力义 *Pu-p'o-li-ch'a.*

The leaves of the two first, and the bark of the last named, arc used for writing on. *Pei-to* is a Sanscrit (梵) word ( patra ), which translated into Chinese signifies "leaf"; *Pei-to-p'o-li-ch'a* ( patra vrikcha ) being translated means "leaf tree." The classics of the Western regions are written on the leaves and the bark of these three kinds of tree, and they may be preserved for five or six hundred years without injury. From *Kiao-chi* ( Cochin China, v. s. ) the wood of this tree has lately been exported as material for the manufacture of bows; for this purpose it answers well.

The Pên-ts'ao describes the same Palm (*XXXI* p. 21), but quotes only the following statement from the *Huan-yü-chi* (close of the 10th century).

緬 甸 *Mien-tien* ( Burmah ) is situated to the South of 滇 *Tien* ( Yün-nan ); it possesses the 樹頭櫻 *Shu-t'ou-tsung* (tree head Palm), which is five or six feet in height and bears a fruit like a Cocoanut; the natives put some leaven (麴) in a jar, which they suspend beneath the fruit, cutting open the fruit so that the liquid runs into the jar; this makes wine which is called "tree head wine:" if leaven be not used they boil the liquid down into sugar. This is the 貝 *Pei* tree. The Burmese use the leaves to write upon.

Finally the History of the Liang dynasty (502-557), Chap. 54, mentions a wine tree, 酒樹 *Tsiu-shu.* From the juice of its flowers wine can be made. This tree grows in 頓遜 *Tun-sun,* a country lying 3000 li to the South of Fu-nan (v. s.) The *Hai-kuo-tu-chi* states, that Tun-sun was in the peninsula of Malacca.

All the above descriptions of Chinese authors point to Palms, the leaves of which are used to write upon and which yield palm wine, and especially to the *Palmyra palm, Borassus flabelliformis.* The Palmyra palm is found throughout India, especially in the dry and hot regions. The limit of its geographical distribution reaches to the North as far as the 25°. It grows in Burmah and may occur also in Yünnan. Grosier (la Chine II. p. 534,) speaks of Borassus tunicata Lour. as of a Chinese palm: "Le Rondier (B. tunicata) croît à la Chine et dans les Indes. Les Chinois méridionaux, comme les Indiens emploient ses grandes et larges feuilles palmées à fabriquer des éventails assez grands pour mettre plusieurs hommes à l'abri du soleil et de la pluie." The fruits of the Palmyra-palm are about the size of a child's head and contain a milky juice like the Cocoa-nut, much used among the natives as medicine. Therefore the ancient botanists called it "nux medica." The long stalked leaves from 8 to 10 feet long, resemble a fan. They are used for many useful purposes, in the manufacture of hats, umbrellas, for thatching roofs &c. The same leaves furnish the paper used by the natives. According to Crawfurd the greatest part of the Pali literature was written on leaves of the Palmyra-palm, from 1 to 1½ feet long, by scratching the letters with an iron stylus. The writings are made legible by rubbing them with a black powder.

The Sanscrit name of the Palmyra-palm is "tala" (rendered by the Chinese sounds *To-lo* v. s.) This name was known by Arrianus, who wrote (second century in his Hist. Ind. VII p. 43: Arborum corticibus Indos vesci solitos fuisse, vocari autem eorum lingua eas arbores Tá-la.

But there is yet another Palm in India the leaves of which supply the natives with paper, the *Corypha umbraculifera,* or *Talipot palm,* a native of Ceylon and the Malabar coast. Some of the sacred books of the Singhalese are writen upon the leaves of this palm.

As regards the *Shu-t'ou-tsung* ( v. s. ) and the mode of obtaining wine from it in Burmah, as described by the ancient Chinese authors, this seems to refer also to Borassus flabelliformis. The "Toddy" or palmwine is obtained from the flower spikes (spathes) of the palm, from which it flows after an incision. It is intoxicating after fermentation. Toddy is also furnished by several other palms of India, namely *Phoenix sylvestris, Cocos nucifera, Arenga saccharifera, Caryota urens.*

### 8 and 9. 鳳尾蕉 *Fêng-wei-tsiao* and 鐵樹果 *Tie-shu-kuo.*

#### *Cycas species.*

I find in the *Chi-wu-ming-shi-t'u-k'ao* the description and engravings of two palms, which are not described separately in other Chinese botanical works. Both seems to refer to species of *Cycas.* The following short accounts are there given of them.

鳳尾蕉 *Fêng-wei-tsiao* (Phoenix' tail's Banana) *Ch. W. XXXVII.* 28.—This is a tree of Southern countries. In Annam it grows abundantly. The trunk is covered with scales. The leaves resemble the leaves of the *Tsung-lü* (v. s.) are pointed, very hard, shining and smooth. If the tree is about to decay it must be burned by a red hot iron nail; then it will thrive again. The Pên-ts'ao identifies the Fêng-wei-tsiao with the Date-palm (v. s.) But the author of the *Ch. W.* believes, that Li-shi-chên is wrong. The drawing in the *Ch. W.* represents a Palm-tree with pinnate leaves.

鐵樹果 *T'ie-shu-kuo* (Fruit of iron tree.) *Ch. W. XXXVI* 43. This tree grows in 滇南 *Tien-nan* (Yün nan province.) On the top of the tree there grows a bundle of leaves, crowded together, which are 7-8 inches in length and resemble in shape a spoon with its handle. From the borders of these spoonlike leaves the fruits proceed. They are roundish, flattened, with a depression in the middle. These fruits are inedible. Within there is a kernel. The natives of Yün-nan call them "Phoenix' eggs." The tree bears fruits only once in every 12 years. It is cultivated in gardens, only as a curiosity, but it is not classed among the fruit trees. The drawing of the T'ie-shu-kuo in the *Ch. W.* represents very well the pinnately cleft fruit-bearing leaves, with the nut like fruits at their margins, so characteristic of the genus Cycas.—What the Chinese tell regarding the revivication of the Fêng-wei-tsiao by iron is practised by the Hindus on the Cycas circinalis. Büsching (Erdbeschreibung, Asien *V.* 4 p. 779) states:

"Merkwürdig ist, dass Cycas circinalis eine grosse Sympathie zum Eisen hat, indem der Baum sogar, wenn er absterben will, durch einen eingeschlagenen eisernen Keil wieder neues Leben erhalten soll."—I. Bontius (Histor. natural Indiae orient. 1631) tells (p. 85): "In Japonia arbor Palmae figura crescit, quae si a pluviis permaduerit, tanquam peste correpta statim exarescit, quam mox cum radicibus avulsam in locum apricum siccandam exponunt indigenae, et tum in eandem scrobem injecta prius arena fervida, aut scoria ferri, replantant, et si qui rami exsicenti, vel avulsi sint vel decidere, eos clavis ferreis trunco affigunt, et sic pristino virori restituitur." This quotation points probably to Cycas revoluta, a Japanese species. This tree, much cultivated in China as an ornamental plant bears at Peking the popular name 鐵樹 *T'ie-shu* * ( iron

---

* The 鐵樹 *T'ie-shu* of Chinese books relates not to a palm, but probably to a species of Dracaena. The description given of it in the Ch. W. XXX 31, is the following:

The T'ie-shu is a little tree, several feet high with an undivided trunk without lateral branches and closely packed joints like a palm. The leaves, which are aggregated at

tree ).—As regards Cycas circinalis, Dr. Hance states (Notes and Queries III p. 95) that there does not seem any evidence of its occurrence on the mainland of China, but it grows wild in Formosa.

These are the palms and palm like trees, the description of which I have been able to find out in Chinese botanical works. But in the Chinese works are left out, I think, some representatives of the Palm order in China. Some European writers mention several species of *Calamus* (Rattans) as growing in Southern China. Grosier (la Chine II 360) states: "Le rotang, que les Chinois appèlent *ten*, croit dans toutes les contrées meridionales de l'Empire; la province de Kouan-ton en fournit une immense quantité, et il abonde surtout dans les environs de *Soui-tcheou-fou*, où les montagnes en sont couvertes. On en distingue plusieurs espèces, dont une se fait surtout remarquer par la prodigieuse longuenr de ses tiges *( Calamus rudentum*. Lour.) L'espèce la plus communne à la Chine, et qu'on emploie à un plus grand nombre d'usages, est celle qui ne pousse qu'une seule tige *( Calamus verus*. Lour.) . Le rotang est très souple et ne se rompt que très difficilement; aussi en tire-t-on le parti le plus utile. Il fournit à la marine Chinoise des câbles et des cordages. On le divise en brins longs et déliés, dont on façonne des corbeilles, des paniers et surtout des nattes, sur les quelles les Chinois couchent en été."

---

the summit, are of a purple colour, resembling the 芭蕉 *Pa-tsiao* (Banana.) Therefore the tree is also called 朱蕉 *Chu-tsiao* (red Banana.) The name T'ie-shu refers to the reddish iron colour of the whole tree. The blossoms resemble those of the 桂 *Kui* (Cinnamomum Cassia.) In Bridgman's Chrestomathy p. 453 the T'ie-shu is identified with *Dracaena ferrea*. In Grosier's "la Chine," III 96, Dracaena ferrea is described as follows: "Cet arbrisseau s'élève à huit pied de haut. Sa tige, d'un pouce de diamètre est simple, à nœuds rapprochés, produits par la chute des feuilles. Il parait appartenir à la famille des palmiers.

Lamarck (Botanique II* p. 324) says regarding *Dracaena terminalis*: "Cette plante s'élève à la hauteur de huit à dix pieds, sur une tige arborée, feuillée à son sommet, et est souvent remarquable par la couleur pourprée que prennent toutes ses parties. Ses feuilles sont grandes, petiolées lancéolées, striées par des nervures latérales, obliques comme dans celles des Balisiers ( Canna). Cette plante croît à la Chine, Burnet states; "Dracaena terminalis is planted as a landmark in China as well as India."

But Dr. Williams (Middle Kingdom I p. 278) says: "The Rattan has been said to be a native of China but this requires proof; all that used at Canton for manufacturing purposes is brought, together with the Betel-nut from Borneo and the Archipelago."

According to Bridgman's Chrestom. the Rattan is called 沙籐 *Sha-t'éng* (sand liana) at Canton. The character *T'éng* corresponds with the European term "liana," for it is used by Chinese writers for many coarse climbing plants. The Kuang-kün-fang-pu (Chap. 81) and also the Pên-t'sao *(Chap. XVIII b,* Twining plants) mention about 50 kinds of *T'éng.* But the Sha-t'éng is not treated of. I cannot find in the *Ch. W.* a drawing, which could be referred to a Rattan.—Dr. Hance observed three kinds of Calamus in the island of Hongkong (Cf. Bentham's Flora Hongkongensis.)

In concluding, I have undertaken to illustrate my notes on Chinese Botany by several Chinese woodcuts, representing plants, treated of in the foregoing paper. They are cut by a Peking artist after drawings from the *Chi-wu-ming-shi-t'u-k'ao* and printed on Chinese paper and according to the Chinese method. Although they do not stand high as specimens of art, they will give at least an idea to the reader of the drawings in the best Chinese pictorial work of this class. I have chosen the following representations.

1. 蜀黍 *Shu-shu.* Sorghum vulgare I. 44.

2. 粱 *Liang.* Setaria italica. I. 18.

3. 薯蕷 *Shu-yü.* Dioscorea Batatas.

4. 檾麻 *T'sing-ma.* Sida tiliaefolia. XIV. 14.

5. 商陸 *Shang-lu.* Phytolacca. XXIV. 3.

6. 佛手柑 *Fo-shou-kan.* Citrus sarcodactylus XXXI. 24.

8. 椰子 *Ye-tsū.* Cocoa-nut. XXXI 18.

7. 鐵樹果 *T'ie-shu-kuo.* Cycas. XXXVI. 43.

---

## ADDENDA.

*Red Rice.*—In treating of the different kinds of Rice known at Peking I omitted to mention a singular variety of rice, called 御稻米 *Yü-tao-mi* (Imperial Rice) or 香稻米 *Siang-tao-mi* (fragrant Rice) or 紅稻米 *Hung-tao-mi* (red Rice.) This Rice is mentioned in the Memoirs of Emperor Kanghi, 1662-1725 ( 聖祖御

製,) quoted in the 《*Shou-shi-t'ung-k'ao,* Chap. 20. The Emperor states, that he once, discovered, while walking among the rice fields in the neighbourhood of his summer palace, a singular rice plant, which was ripe much earlier, than the other rice and bore a very beautiful corn of a red colour and pleasant odour. Kanghi gave orders to have this corn sown in his gardens. Its culture was very successful and this rice was afterwards used for the Imperial table because of its very pleasant taste. As it ripens early it can be cultivated also beyond the great wall (in Mongolia,) where the frost begins very early and ceases very late. The Emperor sent also this rice for cultivation to Che-kiang and Kiang-nan, where two crops yearly can be obtained from it. I am not aware whether the Yü-tao-mi is now generally cultivated in China. But in the neighbour-hood of *Yüan-ming-yüan* (the Imperial summer palace) its cultivation is still continued. The corn of this kind of rice is not completely red, as the Emperor states, but of a pale carnation colour with brown spots. When boiled it becomes very pleasant to the taste.

---

I have expressed some doubt whether *Rye* occurs in the Chinese dominions. Since writing this I read an article of Mr. Simon (Journal of the North China Branch of the Royal Asiatic Society New series No. 4, Carte agricole d. l. Chine), in which he states that Rye is cultivated in the province of Shensi. Mr. S. does not say whether he speaks from his own observation; he does not give the Chinese name of the plant. It was in vain that I looked through Chinese works to make out a cereal, which could be identified with Rye. But perhaps the 黑龍江麥 *Hei-lung-kiang-mai* (wheat from the Amur River,) mentioned in the Memoirs of Emperor Kanghi (quoted in the Shou-shi-t'ung-k'ao, Chap. 26 p. 10) refers to Rye. It is there said that this kind of corn was brought from 鄂羅斯 *Ao-lo-ssu* (Russia). Rye is largely cultivated in Siberia.

---

The *Chinese Oats* 青稞 *T'sing-ko* * in Chinese books is not, as I stated above, identical with our common Oats (Avena sat-

* 'The character 青 *T'sing,* which is met with very often in Chinese descriptions of plants is one of the ambiguous characters, in which the Chinese language is so rich. Morrison translates it by "light green colour," de Guignes by "blue," Schott (Chinese Sprach-

iva), but resemble more the *Pill corn,* (Avena nuda,) the glumes being much shorter than those of Avena sativa and the grain separating very easily from it. The Chinese oats was described by Fischer as Avena chinensis.

I stated above, that at Peking now a days the character 黍 *Shu* is applied to a kind of Panicum, allied to Panicum miliaceum. The corn has glutinous properties and is called 黄米 *Huang-mi* (Yellow corn.) This character *Shu* has been for a long time erroneously used in this connection and this erroneous application of it took place before the 6th century. The Pên-ts'ao *(XXIII 4)* quotes a writer of the 6th century, who states, that the Shu is cultivated to the North of the Yang-tse-kiang. The plant resembles the 蘆 *Lu* (Reed) the corn is greater than millet. The author adds, that this character *Shu* is erroneously applied to another kind of corn 秫. (This character is likewise pronounced *Shu.*) This latter cereal is separately described in the Pên-ts'ao *(XXIII 13.)* The grain called *Huang-mi* is said to possess much glutinous matter. It is used for manufacturing alcoholic drinks. This corn was known to the Chinese in the most ancient times. It seems to me, that the meaning of the character 黍 Shu in ancient times was not glutinous Millet (as Dr. Legge states, c. f. his translation of the Shu-king,) but rather Sorgho, as Dr. Williams translates (Bridgman's Chrestom. p. 449).

I have stated above, that the character 杏 *Sing,* meaning *Apricot* does not occur in the text of the five Cardinal Classics. But Biot in translating the Chou-li states (l. c. I p. 108): "Les paniers de l'offrande des aliments sont remplis avec des Jujubes, des Châtaignes, des Pêches, des Abricots secs &c." Biot translates the character 穋 *Lao* by dried apricots. This is not correct. In the ancient Dictionary Shuo-wen it is explained by 乾梅 *Kan-mei,* dried plums. Cf. also Kanghi's Dictionary.

lehre p. 47) by "bläulich grau, olivenfarbig," Wassilyeff (Chinese Russian Dictionary) by "dark or black." All these sinologues are right, for the character T'sing does not relate to a fixed colour. Its meaning depends upon the thing to which it relates; referring to a horse its meaning is grey, referring to silk it is black, but if it refers to a leaf it must always be translated by dark green.

Regarding the question ventilated above about the native country of the *Ground-nut,* Arachis hypogaea, which Decandolle believes to come from America, I would quote a statement of Piso '(Hist. natur. Indiae occident. 1658 p. 256): "Fructus subterraneus ex oris Africae olim translatus, tandem Americae nativus quasi factus, *Mandobi* vocatur." The further description of this plant and the drawing given of it by Piso without doubt refer to Arachis hypogaea.

It will not be without interest, I think, if I notice here shortly, as an addition to my former statements about *Tea,* the time, when Europeans first became acquainted with this renowned plant. It is well-known, that the use of the Tea was first introduced into Europe by the Dutch East India Company in the first half of the 17th century. But it was described much earlier by European savants. Bontius (Hist. natur. and med. Indiae orient. 1631 p. 87,) gives a very good drawing of the Tea shrub: "De Herba seu Frutice quam Chinenses *The* dicunt, unde potum suum ejusdem nominis conficiunt. B. states, that no European has seen the Chinese Tea plant and that he was indebted for all information about it to the General Spex, who resided several years in Japan and saw it there growing. Tea is first made mention of in the work of Petrus Maffeus (Historiarum Indicarum select. libri *XVI,* 1589, in the 6th and 12th Chap.) I have not seen it, but it is quoted by Bontius.

Having treated in the foregoing notes of the most important cultivated plants of the Chinese and of their origin, it will not appear superfluous if I dedicate also a few words to the *Sugar-cane,* which is extensively cultivated in Southern China,—all the more as the statements of our Savants about the Chinese Sugar-cane do not always agree.

Rondot (Commerce d'Exportation de la Chine 1848, p. 202) states: "La Chine, si nous en croyons les documens historiques des anciens temps, et en juger par les peintures des plus anciennes porcelaines (!), semble être la première contrée qui se soit occupée de la culture de la canne et de l'extraction du sucre."—The same is repeated in Dr. Williams' Commercial Guide, p. 139.

Father Cibot states (Grosier l. c. III, 206): "La canne à sucre ne fut introduite à la Chine que vers la fin du troisième siècle depuis notre ère."

Mr. Stan. Julien notices (Industries de l'Emp. Chinois, p. 204): "La canne à sucre a été introduite en Chine à une époque très reculée mais les Chinois, pendant des longues années, ne surent pas extraire le suc cristallisable du jus sucré. Ce fut dans l'intervalle de temps compris entre les années 766 et 780, sous la dynastie des Thang, qu'un religieux indien, nommé Tseou, voyagant dans la partie occidentale de la province de Ssetchuen, enseigna la fabrication du sucre de canne aux habitants du Céleste Empire."

Let us refer to the Chinese records about the Sugar-cane. I have not been able to find any allusion to the Sugar-cane in the most ancient Chinese works (five Classics). It seems to be mentioned first by the writers of the second century B. C. The first description of it I find in the Nan-fang-ts'ao-mu-ch'uang (4th century) in the following terms.

The 藷蔗 Ché-ché is called also 甘蔗 Kan-ché* (kan, sweet,)or 竿蔗 Kan-ché (kan, a kind of Bamboo.) It grows in Kiao-chi [Cochin China (v. s.)] It is several inches in circumference, several Chang high (1 chang=10 feet) and resembles the Bamboo. The stem, if broken into pieces, is edible and very sweet. The juice expressed from it, is dried in the sun. After several days it changes into Sugar (餳,) which melts in the mouth. This sugar is called 石蜜 Shi-mi (stone honey) by the natives. Ssū-ma-siang-ju (a poet of the second century B. C.) states in his poem Lo-ko, that the sugar-juice possesses the property of removing the bad effects of intoxication. In the year 286 A. D. the realm of Fu-nan (in India beyond the Ganges, v. s. ) sent sugar-cane as tribute. The reader will remark, that here the sugar-cane is not mentioned as indigenous in China.

The Pén-ts'ao gives (XXXIII. 13) a good description of the Sugar-cane and its varieties, of the manufacture of Sugar &c., and quotes several authors of the Liang, T'ang and Sung-dynasties, who describe the plant. In the Kuang-kün-fang-p'u (Chap. 66, p. 17) it is stated, that the Emperor 太宗 Tai-tsung 627-650 sent a man to Mo-ho-to (Magadha an ancient kingdom in India, the modern Bahar) to learn there the method of manufacturing sugar.

The ancient Chinese annals mention often among the productions of India and Persia

the Shi-mi † (stone honey.) This is white crystallized sugar as the Pén-ts'ao explains, called also 白沙糖 Po-sha-t'ang (white sand sugar.) It is hard like a stone and white like snow.

In all probability the Sugar-cane was first cultivated in India, from which locality it spread. There can be found no proof from Chinese sources, that the Sugar-cane passed from China to India, as some authors assert, (Cf. Lindley, Treasury of Botany p. 1003.) The Sugar-cane seems to have been cultivated in India for the making of sugar much earlier, than in China. The Sanscrit name of Sugar "Sarkara" is rendered by Pliny (about our era) by the word "Saccharum," but his statements about sugar are not at all correct, (l. XII. c. 8.) "Saccharum et Arabia fert, sed laudatius India. Est autem mel in arundinibus collectum, gummis modo candidum et fragile amplissimae Nucis Avellanae magnitudine, ad Medicinae tantum usum." The names for Sugar in all European languages are derived from the Sanscrit word Sarkara. The Persian name of Sugar is "kand." This seems to be derived from the Sanscrit "khanda," Sugar in lumps. From the same Sanscrit word is also derived our name Sugar-candy, or crystallized Sugar. The Sugar-cane is largely cultivated in Northern Persia, namely in the province of Mazanderan, near the Caspian sea.—Lindley states (l. c.) that the Venetians first imported the Sugar-cane from India to Europe by the Red Sea prior to 1148.

As regards the cultivation of the Sugar-cane in China now-a-days, the statement of Dr. Williams (Commercial Guide p. 139) is correct, I think, that it is cultivated everywhere South of lat. 30°. But I am astonished to find a statement of Mr. Champion (Industries de l'Emp. Chinois p. 207,) who speaks of the true Sugar-cane as growing in the province of Chili.

The true Sugar-cane (Saccharum officinarum and perhaps other allied species) growing in China, must not however be confounded with what is called the Northern China Sugar-cane. This is the Sorghum Saccharatum, a plant now-a-days largely cultivated in Europe and America for the purpose of manufacturing Sugar from it. This plant was first introduced from Shanghai into France by the French Consul M. Montigny, in the year 1851, whence it spread over Europe and America, after it was proved,

* These names must not be confounded with the 甘藷 Kan-chu (shu), or sweet Potato (v. s.), written with the same characters. The second character however is differently pronounced (Shu according to Kang-hi's Dictionary, Choo according to Morrison) if it refers to the sweet Potato.

† I must here correct an error, into which I fell in stating (Notes and Queries IV. p. 56), that Shi-mi, mentioned as a product of Persia in the Chinese annals, may be the sweet hardened exudation-product of trees.

that it is very rich in Sugar (10-13%). In the year 1862 Mr. Collins was sent from America to China in order to study the mode of manufacturing Sugar from this plant by the Chinese. But he was much astonished at finding, that the Chinese knew nothing about the fact, that Sugar can be obtained from it. The cultivation of it is limited in China. The stem, cut in little pieces is eaten in a raw state. The grain is used like the grain of Sorghum vulgare. In the Chinese botanical works the Sorghum Saccharatum is mentioned under the same name as the Sorghum vulgare. Cf. article 蜀黍 *Shu-shu P. XXIII.* 6, *Ch. W. I.* But it is there said, that two kinds of this plant are cultivated; the one is glutinous and with glutinous Rice is used in manufacturing alcoholic drinks and also made into cakes. This is Sorghum Saccharatum. On account of the glutinous properties of the plant, it is very difficult to obtain Sugar from it in a pure state. The other kind (Sorghum vulgare, or 高粱 *Kao-liang*) is not glutinous. It makes good gruel and cakes and is good for feeding cattle. Cf. Mr. Collin's article regarding the Northern Chinese Sugar-cane in the North China Branch of the Royal Asiatic Society 1865.

In order to complete my notes on Chinese cultivated plants, I ought also to have treated of the *Poppy* (Papaver somniferum,) and now largely cultivated throughout the whole Empire. But this theme has already been largely treated by several writers in our periodicals in China (*Chinese Repository, Notes and Queries &c.*) I will therefore merely remark that the Opium plant is not indigenous in China, but it was brought at the beginning of the 9th century from Arabia. Therefore the first Chinese name for Opium 阿芙蓉 *A-fu-jung (P. XXIII* 24) represents the Arabian name, being "Afyun." Other names, as quoted in the Pên-ts'ao are 鴉片 *Ya-pien* or 阿片 *A-pien*. Both resemble "Opium," which name, as is known, is derived from a Greek word. The popular name of Opium at Peking is 大烟 *Ta-yen* (great smoke.) In the second half of the 17th century the vice of Opium smoking begin to prevail in China. Since England made the Chinese acquainted with the benefit (!) of Opium, they devote a great part of their arable land (illegally however and against repeated Imperial Edicts) to the cultivation of it, and it seems that the Poppy-plant will soon be considered by the miserable Chinese people

of the present day as important a cultivated plant, as the "five kinds of corn" which Emperor Shen-nung, the Father of Agriculture, taught them to sow.

Since writing on the European works, which try to identify Chinese names of plants with European scientific names, I have obtained a small work, treating of the same subject, Essai sur la pharmacie et la matière médicale des Chinois, par Debeaux 1865. I would quote some passages from this treatise, in order to show how useless and unintelligible it is to quote Chinese names of plants in European spelling, without the Chinese characters. It seems to have been unknown to M. D., that before him Tatarinov, Hanbury and others, wrote about Chinese materia medica, for he quotes only as regards the Chinese names Loureiro. Loureiro in his Flora Cochinchinensis gives a good number of indigenous names of plants, but without Chinese characters. I think, these names of Loureiro, quoted by numerous writers on China, as Chinese names of plants, are rather Cochinchinese for it is only in a few cases, that I have succeeded in recognizing them in Chinese botanical works. M. D. gives also a great many new Chinese names. For the most part they are either completely unintelligible, or very distorted, or erroneously applied.

Page 20 M. D. states that *Stillingia Sebifera* is *Ngan-shu* in Chinese, and p. 90: *Pima-tse* ou fruits à peau huileuse (!) nommés aussi *Ho-tien-tse* fruits, qui produisent la lumière (!) sont les graines de l'arbre a Suif, Stillingia Sebifera. But all Chinese and European writers agree, that the Tallow tree is called 烏白木 *Wu-kiu-mu* in Chinese. 蓖麻子 *Pi-ma-tsu* are the seeds of *Ricinus communis.*

Page 69 and 35: Le *Che-tze*, fruit du *Crataegus bibass* resemble par sa forme et sa couleur à une grosse tomate qui serait applatie sur la partie calycinale. M. D. saw evidently the fruits of the 柿子 *Shi-tsu*, *Diospyrus Kaki.—Crataegus bibass* 枇杷 *Pi-pa* in Chinese.

Page 97: *Tsoun*, Allium sativum, *Tsoun-tse*, Allium cepa, according to M. D.—But Allium sativum (*Garlic*) is 蒜 *Suan*, Allium cepa (*Oignou*) 葱 *Ts'ung* in Chinese.

Page 68: *Kiu-hiang*, bois d'Aloes produit par *l'Aloexylou agallochum*, et p. 89: *Tchin-hian*, bois te *Santal jaune.—* But Aloe-wood is

沉香 *Chên-siang*, and the Sandal 檀 香 *T'an-siang.* \*

Page 77. *Pekin-hoa* ou *Man-lan-hoa*, fleurs de *Callistephus sinensis*. The *Chinese Aster* (Callistephus) is called 菊花 *Kü-hua* in Chinese books and this name is known throughout the whole Empire. But there are numerous varieties with different local names.

Page 80. *Lan-hua*, fleurs d' *Olea fragrans*.—Olea fragrans is known to the Chinese as 桂花 *Kui-hua* in Peking as well as in Southern China (Cf. Bridgman's Chrestom. p· 455, Grosier III. p. 22.) But 蘭花 *Lan-hua* is applied to different *Orchideae* (at Peking to *Cymbidium*).

Page 85. *Yen-tchi-hoa*, c'est à dire fleur qui sent la nuit, racines du *Mirabilis Jalapa*. —Mirabilis Jalapa is indeed called 胭脂 花 *Yen-chi-hua* (in Chi-fa, v. Bridgman's Chrest. p. 454) at Canton, but the Chinese characters mean "cosmetic grease."— 夜 來香 *Ye-lai-siang* (fragrancy coming in the night) is *Pergularia odoratissima ( Ch. XXX. p. 13 )*.

Page 87. *Nin-fo-tze*, is *Buck-wheat* according to M. D. As far as I know Buckwheat is 蕎麥 *K'iao-mai* in Chinese books as well as in the popular language throughout the whole Empire (Bridgman's Chrest. p. 447).

Page 89. *Yo-hoan-tze*=*Myristica moshata*. —The only Chinese name for *Nutmeg*, I know, is 肉豆蔲 *Jou-tou-kou* (Cf. Tatarinov l. c. p. 64. Dr. Williams' Commercial Guide).

Page 92. *Lien-tze*, fruits du Châtaigner. The name of the *Chestnut* is 栗子 *Li-tsŭ.*

Page 100. *Tao-ya*=semences d' Orge, Hordeum hexastichon. The Chinese name of *Barley* is *Ta-mai* (v. s).

Page 101 *Kin-tsao-che*, tiges, et semences du *Sorghum Saccharatum*.—Such a name for Sorgho does not exist I think, in China.

\* The Pên-tsao (XXXIV. 28) explains the name *Chên-siang* (Fragrancy sinking under the water) by the heaviness of the wood. Li-shi-chên states, that the Sanscrit name of the wood is 阿迦嚧 *A-kia-nie*. The third character may be a misprint, for the Sanscrit name of Aloe-wood is *Agaru* (Amarakocha l. c. p. 156.)—(Garu=heavy in Sanscrit.

Page 24. La résine d'une espéce de pin, originaire du Thibet est nommé *Pe-go-song* est employé dans tout le Nord de la Chine.— 白果松 *Po-kuo-sung* is at Peking the popular name for *Pinus Bungeana*, a splendid Pine with white bark. It is met with everywhere in the neighbourhood of Peking. As far as I know this tree is not a native of Thibet, and has not been detected elsewhere than in the neighbourhood of Peking.

What M. D. means by *Ku-lo-kiang* (encens mâle) and *Yün-hiang* (encens femelle) page 65, I am not able to state. The Chinese name for Olibanum is 乳香 *Ju-siang.*

Page 93. *Chou-tsao*, tiges feuillées et sommités fleuries du *Cannabis sativa*. Les préparations med. prennent le nom *Houang-yeou*, c'est-à-dire dans le dialecte du Fokien, faisant oublier le chagrin ou la douleur. M. D. believes, that the word *Huang* is derived from the Egyptian or Persian "bengh." These names quoted as Chinese names of Hemp and its preparations, I can nowhere find in the Pên-ts'ao, but his *Houang-yeou* is probably 黃藥 *Huang-yao* (yellow medicine,) whilst 忘憂草 *Wang-yu-ts'ao* (meaning make forgotten sorrow) is given in the *Ch. W. (XIV 42)* as a synonym of 萱草 *Süan-ts'ao*, Hemerocallis graminea according to Tatarinov.

---

## LIST OF CHINESE WORKS, QUOTED IN THE FOREGOING NOTES.

As the greater part of these works cannot be found in Wylie's Notes on Chinese Literature, the information regarding them has been derived from an examination of the 四庫全書 *Ssŭ-k'u-ts'üan-shu*, the great Catalogue of the Imperial Library 1790. I hereby give only the title, the author's names and the time of publication. All these works treating of Materia medica, Botany, Geography, History &c., are often quoted in the Pên-ts'ao and in other Chinese Botanical works.

*Works, written before the third century B. C.*

1. 神農本草經 *Shên-nung-pên-ts'ao-king*. Classical herbal, or Materia medica of the Emperor *Shên-nung*. 2700 B. C.

2. 書經 *Shu-king*. Book of History, compiled by 孔夫子 *K'ung-fu-tsŭ* (Confucius) about 500 B. C. from the historical remains of the time of Emperor Yao (24th century B. C.), the 夏 *Hia* dynasty

(2205-1766 B. C.), the 商 *Shang* dynasty (1766-1122 B. C.) and the 周 *Chou* dynasty, under which Confucius lived.

3. 詩經 *Shi-king* Book of Odes. Collection of ballads used by the people in ancient times in China; also by Confucius.

4. 春秋 *Ch'un-ts'iu.* Spring and Autumn Annals, by Confucius.

5. 周禮 *Chou-li.* Ritual of the Chou dynasty, written in the 12th century B. C.

6. 易經 *Yi-king.* Book of Changes.

*N. B. No. 2-6 are called the* 五經 *Wu-king, the 5 classics.*

7. 爾雅 *Rh-ya.* Literary Expositor, is attributed to 子夏 *Tsu-sia,* a disciple of Confucius ( 5th century B. C.) But a part of it was written in the 12th century B. C.

8. 山海經 *Shan-hai-king.* Hill nd River Classic. It is attributed to the Emperor Yü (2205-2198.)

*Works written during the* 漢 *Han dynasties, 202 B. C.—221. A. D.*

9. 史記 *Shi-ki.* Historical Records by 司馬遷 *Ssu-ma-ts'ien* (second century B. C.)

10. 前漢書 *Ts'ien-han-shu.* History of the anterior Han 202 B. C.—25 A. D.

11. 後漢書 *Hou-han-shu.* History of the posterior Han 25-221 A. D.

12. 三輔黃圖 *San-fu-huang-t'u.* Description of the public buildings in *Changan,* and the Capital during the Han dynasties (second century B. C.)

13. 說文 *Shuo wen,* ancient Chinese Dictionary by 許慎 Sü Shen. A. D. 100.

晉 *Tsin Dynasty 265-420 A. D.*

14. 晉書 *Tsin-shu.* History of the Tsin dynasty.

15. 古今注 *Ku-kin-chu* by 崔豹 *Tsui-pao* (4th century.)

16. 拾遺記 *Shi-yi-ki* by 王嘉 *Wang-kia* (4th century.)

17. 南方草木狀 *Nan-fang-ts'ao-* mu-ch'uang by 稽含 *Ki-han* (4th century.)

18. 吳都賊 *Wu-tu-fu* by 左思 *Tso-ssu.*

19. 廣雅 *Kuang-ya* by 張揖 *Chang-yi.*

魏 *Wei Dynasty* 386-558.

20. 藥錄 *Yao-lu* by 李當之 *Li-tang-chi.*

21. 洛陽伽藍記 *Lo-yang-kia-lan-ki* by 楊衒之 *Yang-sien-chi.* Description of the Buddhist establishments in Lo yang, the Capital of the Wei (beginning of the 5th century.)

梁 *Liang Dynasty.* 502-557.

22. 梁書 *Liang-shu.* History of the Liang.

23. 名醫別錄 *Ming-yi-pie-lu* by 陶弘(宏)景 *Tao-hung-king.*

北齊 *Pei-tse Dynasty* 550-577.

24. 雷公藥對 *Lei-kung-yao-tui* by 徐之才 Sü chi-ts'ai.

唐 *T'ang Dynasty,* 618-907.

25. 唐書 *Tang-shu.* History of the T'ang.

26. 唐本草 *Tang-pên-ts'ao* by 蘇恭 *Su-kung* and 20 other authors. Second half of the 7th century.

27. 海藥本草 *Hai-yao-pên-ts'ao* by 李珣 *Li-sün.* Second half of the 8th century.

28. 本草拾遺 *Pên-ts'ao-shi-yi* by 陳藏器 *Chên-ts'ang-ts'i.* First half of the 8th century.

29. 嶺表錄異 *Ling-piao-lu-yi* by 劉恂 *Liu-sün.*

30. 太平寰宇記 *Tai-ping-huan-yü-ki* by 樂史 *Lo-shi,* a general statistical and descriptive view of the Empire. Close of the 10th century.

31. 吳錄地理志 *Wu-lu-ti-li-chi* by 陸廣微 *Lu-kuang-wei.*

32. 酉陽雜俎 *Yu-yuang-tsa-tsu* Desultory jottings of Yu-yang by 段成式 *Tuang-chêng-shi*, treats of the supernatural and strange, contains much information regarding the productions of China. Close of the 8th century.

宋 *Sung Dynasty* 960–1280.

33. 開寶本草 *Kai-pao-pên-ts'ao* by 馬志 *Ma-chi.* Second half of the 10th century.

34. 圖經本草 *Tu-king-pên-ts'ao* by 蘇頌 *Su-sung.* 11th century (first half).

35. 嘉祐補註本草 *Kia-yu-pu-chu-pên-ts'ao* by 禹錫 *Yü-si* and 林億 *Lin-yi.* First half of the 11th century.

36. 本草衍義 *Pên-ts'ao-yen-yi* by 寇宗奭 *Kou-tsung-shi.* About 1100 A. D.

37. 廣志 *Kuang-chi* by 郭義恭 *Kuo-yi-kung.*

38. 炮炙論 *Pao-chi-lun* by 雷斅(公) by Lei-siao (kung).

元 *Yüan Dynasty,* 1280–1368.

39. 文獻通考 *Wên-sien-t'ung-k'ao,* the celebrated Encyclopaedia of 馬端臨 *Ma-tuan-lin.*

40. 中南志 *Chung-nan-chi* by 黎崱 *Li-tsě.*

明 *Ming Dynasty,* 1368–1644.

41. 本草綱目 *Pên-ts'ao-kang-mu,* the celebrated Materia medica of 李時珍 *Li-shi-chên.* Close of the 16th century.

41. 大明一統志 *Ta-ming-yi-tung-chi.* Geography of the Empire at the time of the Ming.

43. 海槎錄 *Hai-cha-lu,* by 顧玠 *Ku-kie.*

44. 蜀記 *Shu-ki.* Annals of Ssǔ-ch'uan by 曹學 *Ts'ao-sio.*

45. 輟耕錄 *Cho-kang-lu,* by 陶九成(宗儀) *Tao-kiu-chêng (Sung-yi).*

46. 說郛 *Shuo-fu,* by 陶九成 (宗儀) *Tao-kiu-hêng (Tsung-yi).*

大清 *Ta-tsing, the present Dynasty.*

47. 廣羣芳譜 *Kuan-kün-fang-pu.* 1708. vide supra.

48. 授時通考 *Shou-shi-t'ung-k'ao* 1742. vide supra.

49. 植物名實圖考 *Chi-wu-ming-shi-tu-kao.* 1848. vide supra.

50. 大清一統志 *Ta-tsing-yi-t'ung-chi.* Great Geography of the whole Empire of the present Dynasty. Published about the middle of the last century. A new edition issued about 1825.

51. 歷代地理志韻編今釋 *Li-tai-ti-li-chi-yün-pien-kin-shi,* Dictionary of Chinese historical Geography. 1842.

52. 廣東統志 *Kuang-tung-t'ung-chi.* Description of the province of *Kuang-tung.*

53. 廣西統志 *Kuang-si-t'ung-chi.* Description of the province of *Kuang-si.*

54. 貴州統志 *Kui-chou-t'ung-chi.* Description of the province of *Kui-chou.*

55. 雲南統志 *Yün-nan-t'ung-chi.* Description of the province of *Yün-nan.*

56. 四川統志 *Ssǔ-ch'uan-t'ung-chi.* Description of the province of *Ssǔ-ch'uan.*

57. 湖南統志 *Hu-nan-t'ung-chi.* Description of the province of *Hu-nan.*

58. 浙江統志 *Chê-kiang-t'ung-chi.* Description of the province of *Chê-kiang.*

59. 安徽統志 *An-hui-t'ung-chi.* Description of the province of *An-hui.*

60. 南越筆記 *Nan-yüe-pi-ki.* A description of the modern Kuangtung province.

61. 海國圖志 *Hai-kuo-tu-chi* by 魏源 *Wei-yüan.* Historical Geography of foreign countries. 1844.

ERRATA.—The writer of these notes 'not being in the spot while they were passing through the press, several misprints have unfortunately crept in. The more important in the earlier impressions are here noticed, and the remainder are left to the reader.

Page 2, col. 2, (Foot-note) line 17 for There are, read, These are. Page 3, col. 1, line 43, for Paconia, read Paeonia. Page 3, last line, for terrestries, read terrestris. Page 3, col. 2, line 44, for Carror, read Carrot. Page 4, col. 1, line 18, for Tampelmoose, read Pampelmoose. Page 4, col. 2, line 2, for Eroton, read Croton. Page 5, col. 1, line 27, for 粟 read 粟. Page 5, col. 1, line 45, for 'Zifyphus, read Zizyphus. ' Page 5, col. 2, line 6, for Apeciosnm, read Speciosum. Page 9, col. 1, line 3, for Hiang-mi, read Kiang-mi. Page 9, col. 2, line 2, for Fobstears, read Jobstears. Page 10, col. 1, line 27, for Pachyrrhifus, read Pachyrrhizus. Page 10, col. 2, line 27, for Plunus, read Prunus.

梁

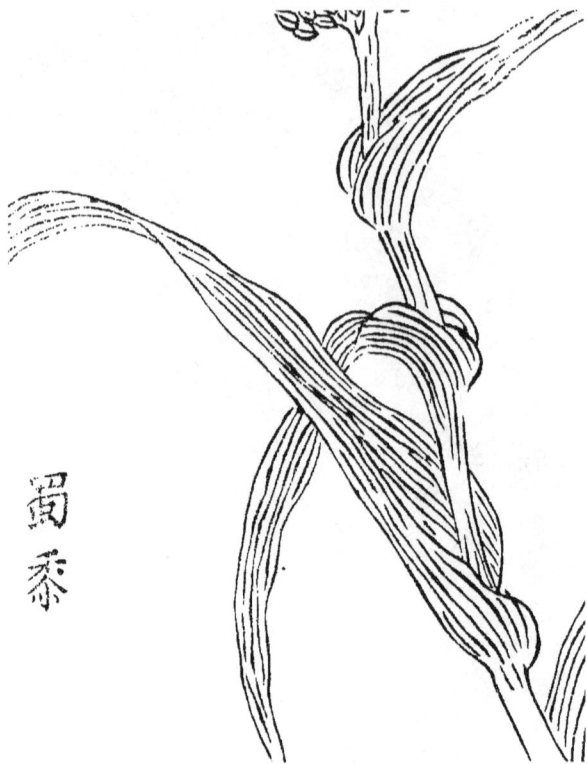

蜀黍

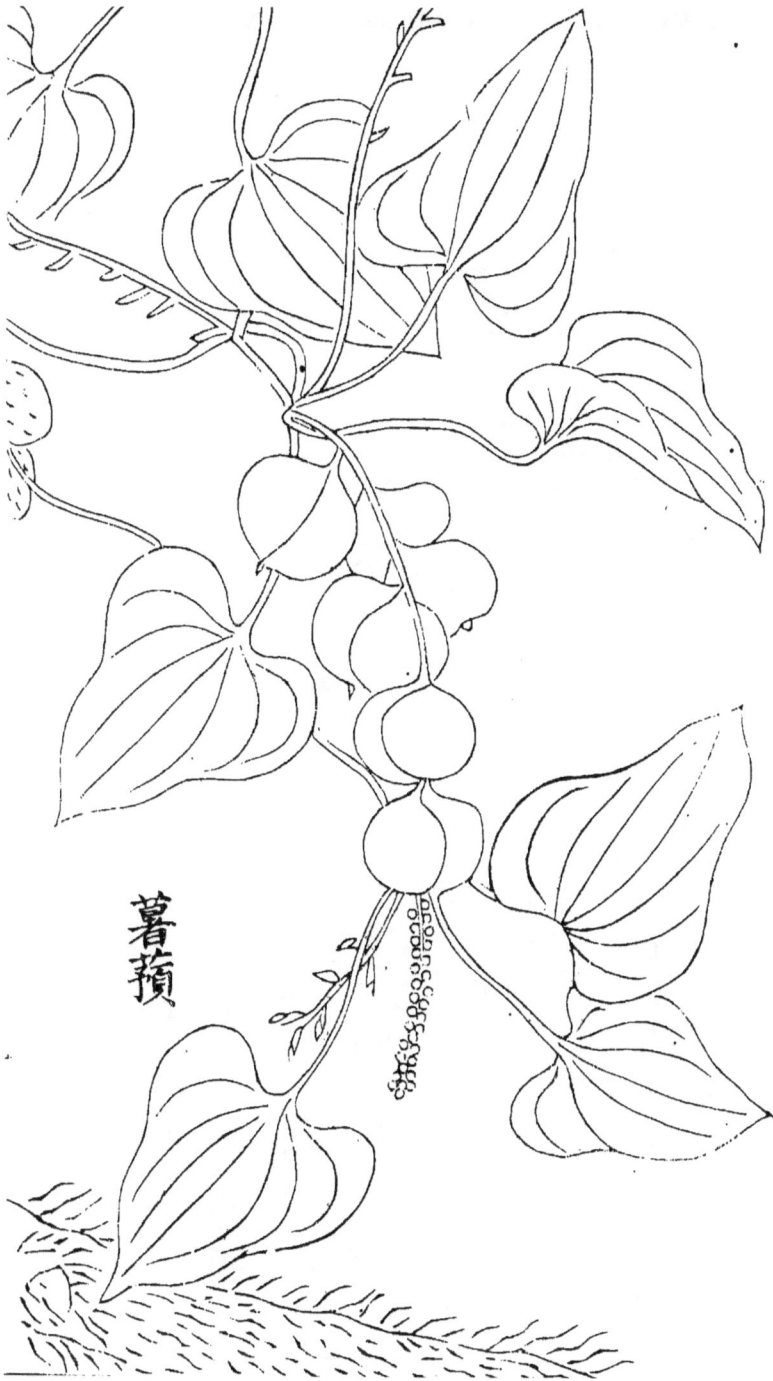

薯蕷

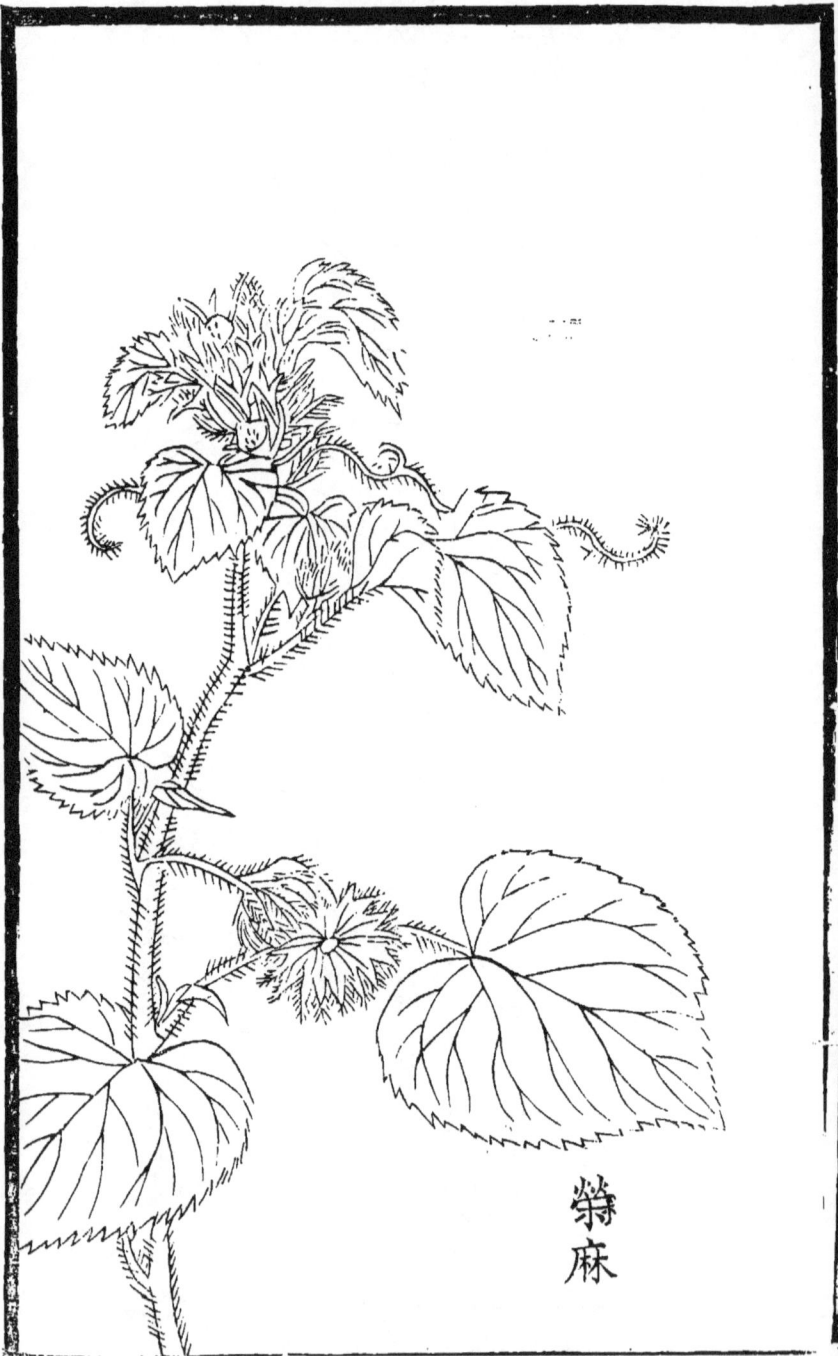

蕁
麻

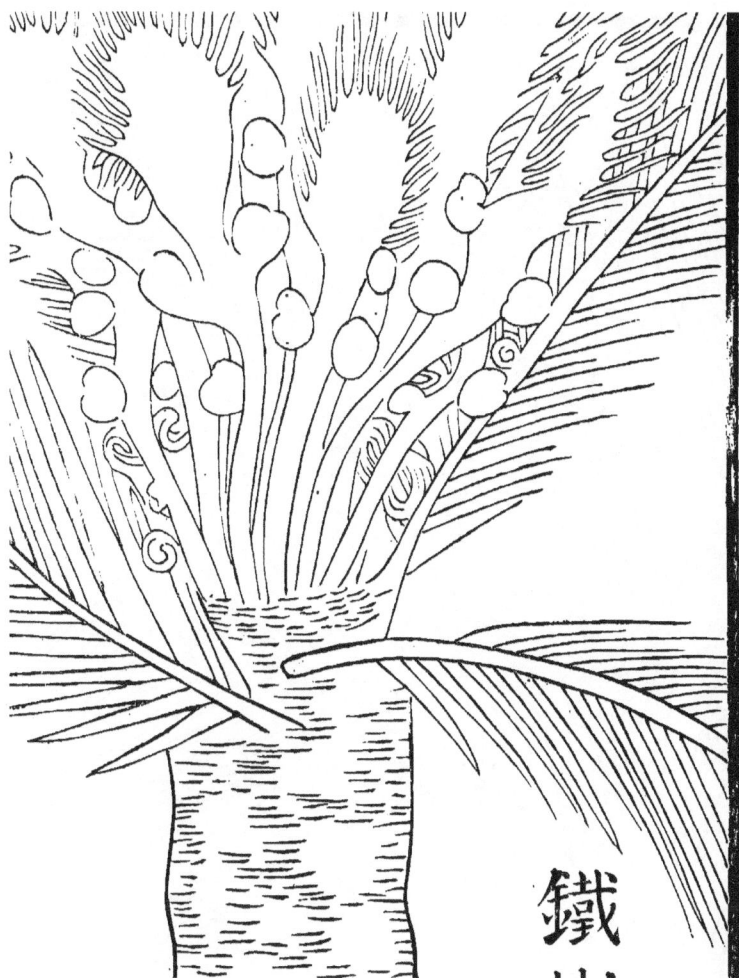

鐵

椰
子

www.ingramcontent.com/pod-product-compliance
Lightning Source LLC
Chambersburg PA
CBHW022010190326
41519CB00010B/1465